Successful Volunteer Retention and Recruitment in the Fire Service

SUCCESSFUL VOLUNTEER RETENTION AND RECRUITMENT IN THE FIRE SERVICE

Dr. Candice McDonald

FIRE ENGINEERING · BOOKS ·

Disclaimer
The recommendations, advice, descriptions, and methods in this book are presented solely for educational purposes. Photos are for instructional purposes only. Always wear the proper level of approved PPE when conducting training drills and operating at incidents. The author and publisher assume no liability whatsoever for any loss or damage that results from the use of any of the material in this book. Use of the material in this book is solely at the risk of the user.

Copyright © 2025 by
Fire Engineering Books
110 S. Hartford Ave., Suite 200
Tulsa, Oklahoma 74120 USA

800.752.9764
+1.918.831.9421
info@fireengineeringbooks.com
www.FireEngineeringBooks.com

Executive Vice President: Eric Schlett
Vice President, Group Publishing: Amanda Champion
Vice President of Content Operations: Starlet Franz
Sales and Customer Service Manager: Lane Nash
Managing Editors: Diane Rothschild and David Rhodes
Production Manager: Tony Quinn
Book Development Editor: Daniel Edward Petrino
Book Designer: Robert Kern, TIPS Publishing Services, Carrboro, NC
Cover Designer: Brandon Ash

Library of Congress Cataloging-in-Publication Data Available on Request

ISBN 9781593706067

All rights reserved. No part of this book may be reproduced, stored in a retrieval system, or transcribed in any form or by any means, electronic or mechanical, including photocopying and recording, without the prior written permission of the publisher.

Printed in the United States of America

1 2 3 4 5   28 27 26 25

To the past, present, and future of emergency services; to the innovative leaders who paved the way for this research; to those with a current passion for creating social change within emergency services; and to the emerging leaders, our future emergency service leaders who will carry this work forward.

This book is also dedicated to my family—Travis, Gage, Madison, and Ray—who have unconditionally supported my love for emergency services, even when it did not love us.

Finally, to all the volunteer firefighters, don't give up.
You are making a difference.

CONTENTS

Foreword by Denis Onieal... ix
Preface... xi

1 Using Data for Recruitment and Retention Efforts.................. 1
2 Understanding the History of the Volunteer Fire Service
   and Fire Department... 9
3 The Value of the Volunteer....................................... 27
4 Find Opportunities to Express Volunteer Value................... 47
5 Understanding the Current Problem with the U.S.
   Volunteer Fire Service... 53
6 Focus on Your Stakeholders....................................... 65
7 Understanding Why Firefighters Disengage........................ 85
8 Becoming a Catalyst for Change in a World of Resistance........ 149
9 Developing a Recruitment and Retention Game Plan............... 161
10 Setting the Course for Organizational Success.................. 167
11 The Power of Organizational Brand and Reputation............... 179
12 Citizen Engagement for Recruitment............................. 195
13 Preparing for Outreach.. 211
14 Incorporating Social Media..................................... 219

15	Volunteering Is a Family Affair	229
16	The Value of Nonwage Benefits	259
17	Recognizing Volunteer Success	269
18	Tracking Your Progress	277
19	Final Thoughts	285

Answers to Chapter Review Questions 287
References 295
Index 307

FOREWORD

Imagine you have a choice to be a passenger in one of three cars going 60 mph in the pouring rain. In the first car, the driver is looking through the windshield a quarter of a mile or so ahead, watching for traffic and road conditions. The driver of the second car is looking through the windshield about 10 feet ahead, also watching for traffic and road conditions. The driver of the third car is driving while looking in the rearview mirror. Your organization and its leaders are the drivers of those cars—the leader looking forward 2 or 3 years into the future, the leader looking to next month, and the leader who wants to return to days gone by, respectively.

That downpour and road conditions are the problem. You're the leader. Which driver are you? And if you don't think your organization is going 60 mph, I have a hot news flash for you: between 2020 and 2022, because of the COVID pandemic, our organizations, communities, and businesses were thrust into the future in a matter of months. Just think about how the workplace, education, health care, and society have changed! You probably have an opinion about how well (or not so well) they've adapted.

This lunge into the future has impacted our beloved fire service in ways we couldn't imagine—changes in the built environment, in technology, in energy generation and storage, in accelerated demands for recruitment and retention, and of course, in the perceptions of the next generation coming up behind us. Of all these issues, the most important and challenging is people. That's why this book is essential—it is a must-read for all who care about the future.

In unambiguous terms, Dr. Candice McDonald has taken her substantial academic research and combined it with her wide-ranging experience and produced a practical road map for the future of the volunteer fire service. It can be used by any organization—it's that good. This is a complete work—combining research, data, case studies, clear examples, and practical exercises into a complete volume that provides a road map to success for every fire service leader.

While I'll admit to being a charter member of the Dr. Candice McDonald fan club, my opinion about this book is based on experience. We're facing a challenging future—the fire service always has. The future is neither good nor bad, but it is coming. How we choose to address the upcoming challenges will determine whether our organizations thrive or wither. It is up to you—not someone else, not the chief, not the city council or the mayor.

It is solely up to you, but this book will help you. Take advantage of the research, study the strategies, apply the examples, and discuss the case studies with the men and women in your organization.

And when you've finished reading this book, don't put it on a shelf—keep it on your desk. You're going to refer to it again and again.

—Dr. Denis Onieal
Deputy U.S. fire administrator (ret.)
Chief, Jersey City (NJ) Fire Department (ret.)

PREFACE

You can either make excuses as a leader and allow your organization to remain status quo, or you can commit to the tough work it takes to change and lead your organization to positive outcomes; the choice is yours.

This is the moment when I challenge you to set aside every excuse why you can't recruit new volunteers and why you can't prevent existing members of your organization from leaving. This is the moment when the shift in mind-set starts and you start your journey forward on a new path.

Reading this book is the first step in broadening your perspectives. This book aims to remove past excuses, offering professionals in fire and emergency services and those involved in their government oversight easy-to-implement strategies to foster first-responder retention and increase recruitment. Many emergency service leaders attend seminar after seminar on recruitment and retention yet never see positive change within their organization. They leave these seminars excited after hearing motivating possibilities but gradually become disappointed again when things within the organization remain the same. The problem is that while these leaders are committed to change within their organization, they lack the practical tools to be the driving change agent necessary to increase retention and create successful recruitment campaigns. It is time to correct that and equip those eager to implement positive change with the tools to do so successfully.

This book was designed to capture best practices and provide fire service leaders with the steps to implement such strategies. These strategies do not discriminate. It does not matter if you are in a small rural department flipping pancakes to purchase bunker gear or if you are in a fully-funded combination department. By reading this book, you are making a commitment to your organization and the future of emergency services. This book is your reference guide to organizational change. It is important to understand that there is no universal solution to solving recruitment and retention issues; the specific issues tied to each locality must be addressed.

This book is based on two central ideas. The first is that to be successful, leaders must base their actions on the interest of their stakeholders over self-serving agendas. In case the term is new to you, *stakeholders* are those internal or external individuals or groups who can have an impact on an organization's success. Simply put, your stakeholders are *your* people—both within your organization and outside of the organization that you serve. The second idea is that the responsibility to create change within the organization starts at the top. Leaders with a desire to see their organization succeed will be open to equipping themselves with the tools necessary to drive change. A successful leader will make a personal investment to gain the skills, knowledge, and resources needed to motivate others.

This book originated as the result of my frustration with the negative talk regarding the decline of volunteers in the fire service. The foundation for this writing comes from my doctoral study. A qualitative methodology was used in my study to explore the issue of the need for strategies in retaining volunteer firefighters. The need for this book was sparked by the feedback that I received after workshops on recruitment and retention that I delivered in four different countries. Emergency service leaders from all over the world expressed a strong desire for help. They not only wanted to be provided with resources but also sought an understanding of why recruitment and retention challenges exist.

Prior to my research, academic literature was lacking on strategies for volunteer fire service leaders to prevent volunteer turnover. Those few strategies offered were not accessible to or understandable by the everyday firefighter. They were often locked behind academic journal paywalls. The previous research that was accessible did not offer specific strategies that emergency service leaders could easily adopt. I quickly learned that the overall issues were that volunteer fire service leaders lacked access to the necessary information to help them to identify the problems leading to firefighter turnover and they lacked affordable strategies to fix these problems. This book is intended to make understandable to readers the contributing factors leading to poor volunteer retention and offer research-based strategies to increase recruitment and retention.

While the basis of this book is my research and real-world experience, many others deserve to share the credit. Many dedicated fire service professionals have pioneered new recruitment and retention tactics and have positively impacted the industry. Their groundbreaking efforts will be highlighted in this book. Importantly, the strategies in this book will continue to evolve as changes in the fire service, as well as other changes in the external environment, create new demands for stakeholders.

Each chapter of this book offers easy-to-implement and low- to no-cost strategies. Remember: there is no one-and-done solution to solving the recruitment and retention problem facing the volunteer fire service. Instead what is required is a continuous process that requires an ongoing commitment.

1
USING DATA FOR RECRUITMENT AND RETENTION EFFORTS

Trying to move a fire department forward during high turnover is like trying to finish a race with a leaking tire. You will eventually get to the finish line, but your team will be exhausted from constantly needing to pull over and put air in the tires, and you most likely will come in dead last after exhausting all your resources.

It is no secret: The general community wants to know without a doubt that when they dial 9-1-1, professional and trained fire service and emergency medical service (EMS) personnel will respond to provide emergency assistance. The big question is, Who can make that happen? History and data tell us that volunteers *can* meet this demand.

Of the total estimated 1,063,400 firefighters across the United States,[1] 676,900 are volunteers (fig. 1–1).[2] Citizens serving as fire service and EMS personnel come from diverse backgrounds and donate their time 365 days a year to protect their fellow neighbors. In most areas, the people being served by volunteer first responders can't distinguish between those responding as volunteers and career personnel.

Many fire service leaders are so focused on the decline in volunteers that they fail to embrace the positive data: Almost 700,000 Americans are volunteering to serve as members of volunteer fire departments. No other volunteer organization comes close to the number of people and hours donated within the volunteer fire service. That is something to be proud of—something we should be sharing.

I often hear the excuse that the fire service is different—it is a noble cause that impacts the lives being served. Some say we are unlike other volunteer organizations, and other volunteer organizations don't require the extensive training the fire service does. Volunteers outside of the fire service are not subjected to *compassion fatigue*, a form of burnout resulting from constant exposure to trauma.

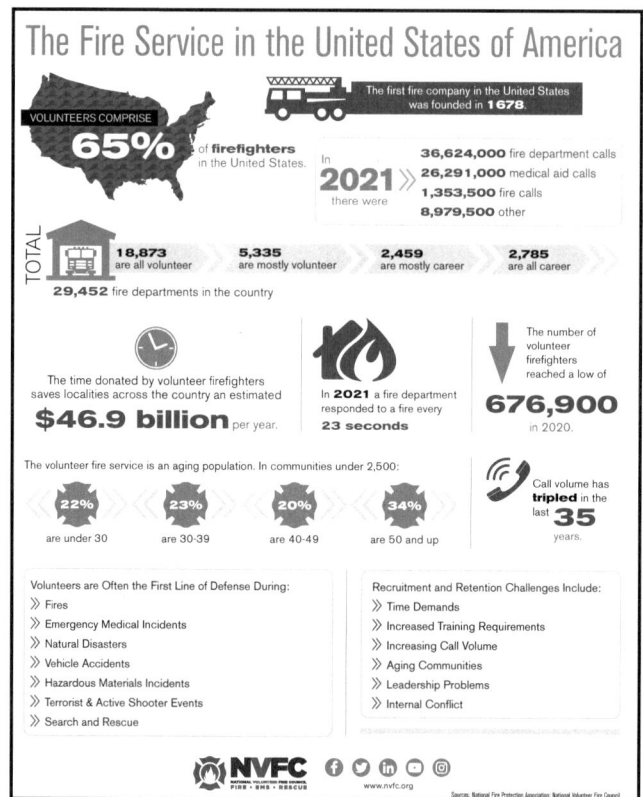

Figure 1–1. Infographic showing the breakdown of volunteer firefighters in the United States from 2021

Compassion fatigue, also called *secondary traumatic stress disorder*, is the disruptive by-product of serving a traumatized population.[3]

Data tells us a different story. We are not alone. Before joining the volunteer fire service, I served as a volunteer court-appointed special advocate (CASA) for abused and neglected children. On average, a child enters foster care every 45 seconds in the United States, and there are only 93,000 CASA volunteers across the country.[4] Contributors to CASA high turnover may include the lengthy and complicated process to become a volunteer, restrictive laws and regulations, limited outcomes for those being served, and unrealistic expectations for volunteers.[5] Do all of those reasons for lack of volunteers sound familiar? The volunteer fire service has seven times the number of volunteers serving across the country as the CASA program, so why are we so focused on the decline over the past few decades? Why are we not telling a story using the data that demonstrates the existing commitment our country has for the volunteer fire service? Numbers are our friend, so let's use the data to tell the story.

The Value of Data in the Volunteer Fire Service

The term *data* scares some leaders, as they equate data with mathematics. If you are one of those people who cringes at the thought of using data, I promise this is simple. Data is our friend. Data-driven approaches enable fire service leaders to make informed decisions, enhance recruitment and retention efforts, and create supportive environments.

Data can be a highly valuable resource for the successful recruitment and retention of volunteer firefighters for the following reasons:

- *Understanding needs.* Data can help fire service leaders to identify needed coverage areas, gaps in services, and times when additional volunteers are needed. This then allows fire departments to target their recruitment efforts more effectively to meet the needs of the community they serve.
- *Analyzing volunteer demographics.* By analyzing demographic data on existing volunteers, fire departments can gain insights into the backgrounds of their existing workforce. This information can inform targeted recruitment strategies to attract a diverse and representative group of volunteers to serve the community.
- *Performance evaluation.* Data can be used by fire service leaders to assess the performance and contributions of existing volunteers. This leads to recognizing the efforts of volunteers and providing constructive feedback to improve the skills of the workforce.
- *Resource allocation.* Data on volunteer availability and response rates can aid in allocating resources efficiently, ensuring appropriate coverage of the community and ensuring that response times are compliant with National Fire Protection Association (NFPA) standards.
- *Tracking retention rates.* Data can help fire service leaders to track retention rates among their volunteers, allowing departments to address issues that may lead to attrition and implement mitigation strategies.
- *Improving training programs.* Fire service leaders can analyze data on training attendance, completion, and competency levels to enhance their training programs, ensure that volunteers are adequately prepared for their roles, and determine the accessibility of training to volunteers.
- *Creating support systems.* Fire service leaders can use data to identify areas where additional support could be needed, such as behavioral health resources, training workshops, or mentoring programs.

Data as a Framework for Storytelling

Did you know that firefighters respond to a call every 23 seconds in the United States?[6] The Any Town Fire Department responded to 322 calls last year, and we have an exciting volunteer opportunity opening soon that *you* would be perfect for! We are looking for six new volunteer firefighters and three new support volunteers. Don't worry—training is provided! Everyone who joins our organization has a great chance to make an impact within our community. There might not be pay, but we do have cool hats and shiny red trucks you can drive! We believe Any Town Fire Department is an awesome place to volunteer, so click here to learn more!

Think about the types of data accessible to you and how data can help tell your story. Combined with a narrative, data can tell a story that bolsters your recruitment and retention efforts. *Data-driven storytelling* is the process of transforming analytical data into visual forms that can influence the decisions of stakeholders. Fire service leaders can leverage data-driven storytelling to share their needs and actively generate change.

The three main elements of data-driven storytelling are data, visuals, and narrative. As you are developing your data story, consider your audience, know who the key players are, identify the problem, and think about what you are trying to explain, what your goals are, what problem you are hoping to solve, and what outcome you hope to achieve with the help of the audience.

Compelling data-driven storytelling should include:

- Visuals that represent the evidence behind your point
- Relevant information only, to avoid distraction
- Timely data that highlights current trends only
- Ethical data to ensure that you are not misleading the audience to get an outcome
- Trends to show what is staying the same, changing, or developing
- Correlations between data points to show connections

Data-driven storytelling, when used correctly, is a powerful tool for recruiting volunteers by presenting compelling narratives backed by data. By combining data and storytelling, fire departments can create compelling narratives that resonate with potential volunteers, external stakeholders, and funders; that drive recruitment efforts; and that foster retention.

Effective data-driven storytelling will achieve the following:

- *Highlight impact.* Fire service leaders can use data to showcase the positive impact that volunteers have made in the community and

the fire department. For example, fire departments can share statistics on the number of community members served, number of hours volunteered, and success stories that demonstrate the difference specific volunteers have made.
- *Show growth and need.* Fire service leaders can present data that illustrates the growth of the fire department and the increasing demands being placed on volunteers. This information is essential when asking for increased funding for nonwage benefits, equipment, training, and personal protective equipment.
- *Personalize stories.* Fire service leaders can use data to personalize stories of current volunteers, by sharing their backgrounds and experiences, as well as conveying the difference that they have made while serving as a valued member of the department. This can help potential recruits see themselves as part of the fire department and understand the meaningful contributions that they can make.
- *Visualize volunteer and department success.* Fire service leaders can use data visualizations such as graphs, charts, or infographics to showcase the progress and achievements of the fire department and its members. Visual representations can make the impact more tangible and easier to comprehend.
- *Address stakeholder concerns.* Fire service leaders can use data-driven storytelling to address common concerns of potential volunteers. For instance, if time commitment is an issue, then show data on flexible scheduling options to meet department requirements.
- *Demonstrate department transparency.* Fire service leaders can use data to be transparent about the fire department's goals, as well as barriers and how they are being addressed, to build trust with potential recruits and stakeholders. Presenting such data openly shows that the fire department has a culture of accountability and responsibility.
- *Appeal to emotions.* Fire service leaders can pair data with emotional storytelling. Combining the numbers with heartwarming anecdotes, testimonials, and personal journeys can evoke empathy and a sense of purpose, which may lead to action among external stakeholders.
- *Incorporate social media.* Fire service leaders can use social media platforms for data-driven storytelling. Sharing visually appealing posts, short videos, and infographics that communicate the impact of volunteering in a concise and engaging way may appeal to stakeholders.

- *Track volunteer progress.* Fire service leaders can use data-driven storytelling after recruitment to update volunteers on the progress and impact of their efforts. This can be done by regularly sharing success stories and metrics internally to keep volunteers motivated and engaged.

CASE STUDY: DATA-DRIVEN STORYTELLING

Pro Football Hall of Fame Pitch

As the deputy chief executive officer of the National Volunteer Fire Council (NVFC), I had a big idea: to approach the Pro Football Hall of Fame (HOF) and ask them to host a Volunteer Firefighter Awareness Day to promote recruitment, educate the public, and honor volunteers. I used national and local data to support my narrative of why they should engage with this idea. I came equipped with multiple copies of both Ohio's and the NVFC's fact sheets on volunteer firefighters (fig. 1–2).

These were my driving points:

- About 70% of Ohio's and the nation's fire service is made up of volunteers.[7]
- Volunteer firefighters save their communities $46.9 billion.[8]
- The volunteer fire service is in desperate need of new volunteers to continue to provide the economic cost savings it has been providing since the days of Ben Franklin.
- The HOF can promote the spirit of volunteerism by partnering with the NVFC to host a Volunteer Firefighter Awareness Day.
- The proposed event has the ability to attract firefighters and their families from all over the country.

That meeting went so well that before I left the HOF asked us to hold potential dates on both our calendars as we worked out the details.

Thoughts to Consider

- What data do you think the HOF found useful in my pitch?
- What opportunities exist for the HOF if they move forward with this event?
- Why was it important to print out and give the HOF the Ohio and NVFC volunteer firefighter fact sheets?

1 | Using Data for Recruitment and Retention Efforts

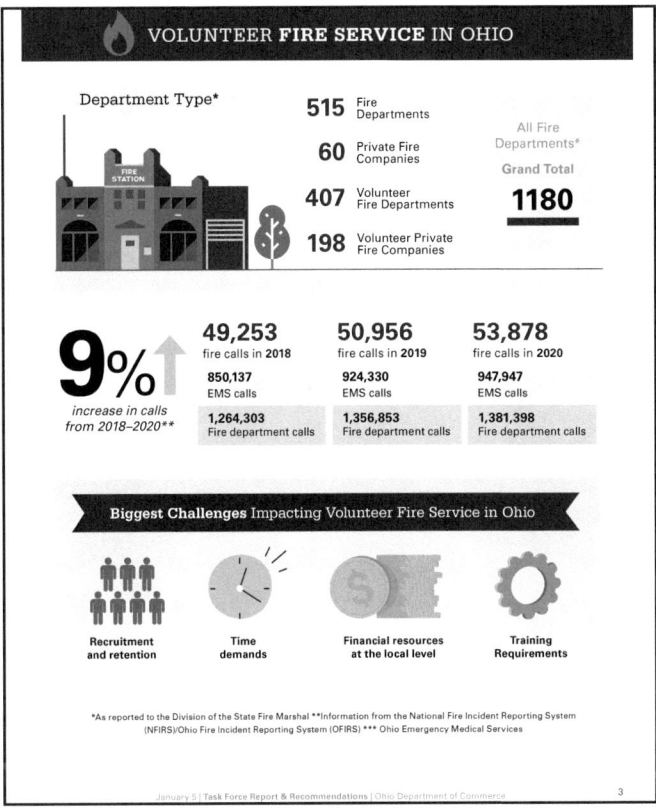

Figure 1–2. Volunteer firefighter fact sheet for Ohio

Review Questions

1. What is data-driven storytelling?
2. List at least three benefits of data-driven storytelling.
3. Compelling data storytelling should include visuals that represent the _____ supporting your point.

Discussion Questions

1. How is your department currently using data?
2. How does your department track and organize data?

2

UNDERSTANDING THE HISTORY OF THE VOLUNTEER FIRE SERVICE AND FIRE DEPARTMENT

Before the beginning of our country to modern times, the volunteer fire service has had a profound impact on the well-being of society. For many early Americans, the first place they got to exercise a right to vote was in their volunteer fire company elections of officers. This is because, before independence, the British Crown appointed all government leaders—governors, mayors, etc. Even today, all these many decades later, many local public election polling places are located in volunteer firehouses. In fact, some of the very same collapsible voting booths used in those earliest elections were utilized both by the general public on election day and at other times by volunteer fire personnel during fire company elections, too. And, our country's first five presidents were affiliated in one or more ways with their local volunteer fire departments.

—Wayne Powell, fire service historian

Using History to Pave the Road to the Future

History can provide fire service leaders with valuable insights that can inform and shape how they plan recruitment and retention efforts. By learning from the past, fire service leaders can make informed decisions, build on past successes, and avoid repeating mistakes.

Understanding the history of the fire service is important for fire service leaders as they plan their future recruitment and retention goals for the following reasons:

- *Learning from past mistakes.* Fire service leaders and members who study history can learn from the mistakes and failures of the past. They can get a sense of what didn't work and why. By

understanding what went wrong previously, fire service leaders can avoid repeating the same errors and investing resources in the wrong place. Are you asking for candid feedback and tracking responses on the outcomes of specific recruitment and retention efforts?
- *Identifying successful recruitment and retention strategies.* History can provide fire service leaders with insights into successful recruitment and retention strategies that were used effectively in the past. Analyzing past successes can inform fire service leaders as they conduct their future recruitment and retention planning. Informed decision-making in turn will lead to better outcomes. Are you tracking what outcomes came from specific recruitment and retention activities?
- *Recognizing patterns and trends.* History can reveal to fire service leaders the patterns around recruitment and retention that can help them to anticipate future trends. Are you noticing patterns with turnover among a specific generation, lower engagement during a certain time of year, or turnover happening repeatedly at a particular time in service?
- *Cultural and social context.* History can help fire service leaders to understand the cultural, social, and political context of different eras. This knowledge is beneficial for planning initiatives that align with the values and needs of the community. How has the community culture changed since you implemented your policies, procedures, and activities? Do you need to make updates to meet the shift in times?
- *Preserving knowledge and traditions.* History offers a foundation of knowledge and experiences that fire service leaders can use to create innovative and effective solutions for the future. Studying the history of the fire service industry and of your individual organization ensures that valuable knowledge, traditions, and practices are preserved in recruitment and retention planning. It also can help fire service leaders to recognize areas for change, such as that hazing is no longer acceptable. This will ensure that those traditions that are critical and beneficial are passed on to future generations. What traditions are important to keep within your organization? What traditions are no longer of value and need to be updated?
- *Fostering critical thinking.* Fire service leaders can analyze past events and outcomes to encourage critical thinking. This also allows them to evaluate different perspectives, collect qualitative and

quantitative information, and make well-informed decisions for the future. Do you solicit feedback from all members, not just leaders?
- *Long-term perspective.* Understanding historical trends can help fire service leaders to adapt and plan for changing circumstances and uncertainties in the future. History can provide fire service leaders with a long-term perspective. This will enable them to see how actions and decisions can have lasting impacts and reinforce the need for well-thought-out plans that are sustainable. What is your long-term succession plan? Who will you need to develop for future advancement among the leadership ranks?
- *Cultural identity and heritage.* Positive and negative history have an impact on a community's cultural identity and heritage. Understanding this heritage is vital for fire service leaders as they create plans to reach into the communities. Is the heritage of the fire department documented and discussed? How can we avoid repeating past mistakes, conflicts, and tragedies?

U.S. Fire Service History

Many fire service leaders are unfamiliar with the true history of the volunteer fire service. Knowledge of the history can be drawn from to provide context during conversations and presentations about the volunteer fire service. For example, did you know that Ben Franklin founded the volunteer fire service in 1736 and you can be part of a nearly 3-century tradition?

The United States has a rich history of volunteer firefighters, with many fire departments having multiple generations of members from the same family. It has been said that the volunteer fire service was officially started almost 300 years ago, in 1736, when Ben Franklin founded the Union Fire Company in Philadelphia.[1] The National Archives cites the Articles of the Union Fire Company as having been written on December 7, 1736.

ARTICLES OF THE UNION FIRE COMPANY[2]

1. That we will each of us at his own proper Charge provide Two Leathern Buckets, and Four Baggs of good Oznabrigs or wider Linnen, Whereof each Bagg shall contain four Yards at least, and shall have a running Cord near the Mouth; Which said Buckets and Baggs shall be marked with the Initial Letters of our respective Names and Company Thus [*A.B. & Company*] and shall be apply'd to

no other Use than for preserving our Goods and Effects in Case of Fire as aforesaid.
2. That if any one of us shall fail to provide and keep his Buckets and Baggs as aforesaid he shall forfeit and pay unto the Clerk for the Time being, for the Use of the Company, the Sum of *Five Shillings* for every Bucket and Bagg wanting.
3. That if any of the Buckets or Baggs aforesaid shall be lost or damaged at any Fire aforesaid The same shall be supplied and repaired at the Charge of the whole Company.
4. That we will all of us, upon hearing of Fire breaking out at or near any of our Dwelling Houses, immediately repair to the same with all our Buckets and Baggs, and there employ our best Endeavours to preserve the Goods and Effects of such of us as shall be in Danger by Packing the same into our Baggs: And if more than one of us shall be in Danger at the same time, we will divide our selves as near as may be to be equally helpful. And to prevent suspicious Persons from coming into, or carrying any Goods out of, any such House, Two of our Number shall constantly attend at the Doors until all the Goods and Effects that can be saved shall be secured in our Baggs, and carried to some safe Place, to be appointed by such of our Company as shall be present, Where one or more of us shall attend them 'till they can be conveniently delivered to, or secured for, the Owner.
5. That we will meet together in the Ev'ning of the last Second Day of the Week commonly called Monday, in every Month, at some convenient Place and Time to be appointed at each Meeting, to consider of what may be further useful in the Premises; And whatsoever shall be expended at every Meeting aforesaid shall be paid by the Members met. And if any Member shall neglect to meet as aforesaid, he shall forfeit and pay the Sum of *One Shilling*.
6. That we will each of us in our Turns, according to the Order of our Subscriptions, serve the Company as Clerk for the Space of one Month, Viz. That Member whose Name is hereunto first subscribed shall serve first, and so on to the last, Whose Business shall be to inspect the Condition of each of our Buckets and Baggs, and to make Report thereof at every Monthly Meeting aforesaid, To collect all the Fines and Forfeitures accruing by Virtue hereof; To warn every Member of the Time and Place of Meeting, at least Six Hours before Hand. And if any new Member be proposed to be admitted, or any Alteration to be made in any of the present Articles, he shall inform every Member thereof at the Time of Warning a[foresaid.] And shall also read over a Copy of these Presents, and a List of all the

Subscribers Names, at every Monthly Meeting, before the Company proceeds to any other Bu[siness,] Which said Clerk shall be accountable to the Rest of the Company for, and pay [to] the next succeeding Clerk, all the Monies accruing or belonging unto the Company by virtue of these presents. And if any Member shall refuse to serve as Clerk in his Turn aforesaid, he shall forfeit the Sum of *Five Shillings*.

7. That our Company shall not exceed the Number of twenty-five Persons a[t a] time, no new Member be admitted, nor any Alteration made in these present Ar[ticles] until the Monthly Meeting after the same is first proposed, and the whole Company acquainted therewith by the Clerk as aforesaid; Nor without the Consent of Three Fourths of our whole Number, the whole Three Fourths being met. But the other Affairs relating to the Company shall be determined by Three Fourths of Members met. And that the Time of entring upon Business shall be one Hour after the Time appointed for Meeting as aforesaid.
8. That each Member shall keep two Lists of all the Subscribers Names, [one] to be fixed in open View near the Buckets and Baggs, and the Other to be pr[oduced] at every Monthly Meeting if required under pain of forfeiting the Sum of [six?] *Pence*.
9. That all Fines and Forfeitures arising by Virtue hereof, shall be paid unto the Clerk for the Time being, for the Use of the Company, and shall be erected into a common Stock. And if any Member shall refuse to pay any Fine or Forfeiture aforesaid when due, his Name shall be razed out, And he shall from thenceforth be excluded the Company.
10. Lastly that upon the Death of any of our Company, the Survivors shall in time of Danger as aforesaid, be aiding to the Widow of such Decedent during her Widowhood, in the same Manner as if her Husband had been living; she only keeping the Buckets and Baggs as aforesaid. In Witness whereof we have hereunto set our Hands; Dated the Day and Year abovesaid.

While the fire service has come a long way from the original horse-drawn wagons, our desire to serve remains the same. Despite what some may say, the data supports that people still want to volunteer for the causes they care about and be empowered to make a difference in the communities they serve. Serving as a volunteer offers an avenue for an individual to connect to their community. The more connected to a community one feels, the greater will be the sense of pride and responsibility to the community that individual has.

The desire to volunteer and make a difference becomes even greater during times of tragedy. After 9/11, the American Red Cross was overwhelmed with people ready to donate their time and compassion to others while still trying to deal with the tragedy themselves. In the hours after the attack, the Red Cross reported involving teams of more than 57,000 volunteers, many of whom had just joined in response to the tragedy.[3]

History demonstrates the desire people have for serving others, especially in times of great need. How have we tapped into the desire to serve after local, state, and national tragedies? When a house fire happens within the community, do we just say we responded, or do we say we responded *and* if you are interested in helping your fellow neighbor, here is how? Are we prepared to offer our community members a way to help their neighbors when something major happens? Have we communicated the available opportunities to volunteer, both emergency and nonemergency?

Establishment and Impact of State and National Volunteer Fire Service Organizations

Do you know—or have you ever wondered—how and why your state firefighters association was formed? Have you ever asked why we have so many national fire service organizations? Most likely, your answer may mention that there is power in numbers and an identified need for collective change.

As the fire service evolved, national fire service organizations began to form, and to this day they play a vital role in advocacy. These interest organizations have continued to grow, gaining strength in membership numbers, increasing the resources that they offer, and offering their expertise to address complex and significant issues of the fire service at the state or national level. These groups have a strong history of influencing public policy and making a lasting impact on society.

Engaging with state and national interest groups can have many benefits for a local fire service leader. As these groups seek to influence public policy, it is important for local volunteer fire service leaders to be aware of issues being brought forward and to provide input on the impact of such issues at a local level.

Collective Voice

Interest groups can provide a collective voice for volunteer firefighters. State and national fire service organizations have the ability to bring together a large number of individuals or groups of stakeholders with a shared interest in the betterment of the fire service. By uniting under one umbrella, the membership can leverage their collective voice to have a greater impact on policymakers and the public.

For example, in 1989 the Congressional Fire Services Institute (CFSI) was formed to make the U.S. Congress aware of the concerns of fire and emergency services. CFSI has provided both career and volunteer firefighters a collective voice on key issues. This includes advocating to elected officials to support the reauthorization of the Assistance to Firefighters Grant Program (AFG), the Staffing for Adequate Fire and Emergency Response Grant Program (SAFER), and the U.S. Fire Administration (USFA).

How does this impact your local recruitment and retention efforts? AFG provides funding to volunteer fire departments to enhance the health and safety of first responders and to improve abilities in responding to fire and fire-related hazards. Many volunteer departments rely on AFG to outfit new and existing volunteers with necessary personal protective equipment.

The funding priorities and criteria for evaluating AFG applications are established by the Federal Emergency Management Agency (FEMA) based on the recommendations from the Criteria Development Panel (CDP), which is made up of nine interest groups:

- Congressional Fire Service Institute (CFSI)
- International Association of Arson Investigators
- International Association of Fire Chiefs
- International Association of Fire Fighters (IAFF)
- International Society of Fire Service Instructors
- National Association of State Fire Marshals
- National Fire Protection Association (NFPA)
- National Volunteer Fire Council (NVFC)
- North American Fire Training Directors

These nine interest groups are responsible for determining—creating new or modifying previously established—funding priorities, as well as developing criteria for awarding the grants that local departments apply for. These same groups appoint peer reviewers to score all the grants received. They make the determination if you get funded or not.

How is your local voice being heard at the state and national levels? Are you a member of any of these special interest groups? Do you provide feedback when they send requests for input?

Increased Influence

National fire service organizations have a broader reach and influence than do local or regional groups. They are often invited by government agencies, such as the USFA and the Department of Transportation, to have a seat at the table and take part in national conversations to provide insight as decisions are made that impact volunteers. They can influence legislation, policies, and public opinion on a larger scale.

The 2023 U.S. Fire Administrator's Summit on Fire Prevention led to the development of Recruitment and Retention Work Group, to "Invest in recruitment and retention initiatives and incentives to address the shortage of firefighters and make the fire service more diverse and inclusive."[4] This work group is guiding the nation's recruitment and retention efforts and consists of members of the aforementioned special interest groups. Becoming active in one of these national groups may eventually provide the local fire service leader with an opportunity to serve on one of these national initiatives. Are you aware of the recommendations this work group is making that will impact you on a local level?

Resource Pooling

By pooling resources from various fire service members or stakeholders regionally, on a state level, or nationwide, fire service organizations can access significantly more grant funding, corporate sponsors, expertise, and strength to support advocacy efforts effectively. Many state organizations are applying for state SAFER grants to address recruitment and retention programs for the stakeholders within their state.

An example of this is the Firefighters Association of the State of New York (FASNY). FASNY successfully received a SAFER grant to provide local fire departments with resources including the Higher Education Learning Plan (HELP). Under the HELP program, FASNY offers student-volunteers a 100% tuition reimbursement (with a cap of $1,500 per semester) in exchange for volunteering with any of New York's volunteer fire departments. Another FASNY resource is the statewide recruitment program RecruitNY, with campaign materials for departments to download (at www.recruitny.org/resources/) (fig. 2–1).

Have you explored what resources your state association offers? Is there a way to partner with other departments within your area to apply for a regional grant to address barriers to recruitment and retention? Could you partner with neighboring departments to host a weekend recruitment event and pool resources to develop campaign materials?

Consistency and Coordination

National fire service organizations can ensure consistency and coordination in advocacy efforts. These organizations have the ability to develop a cohesive strategy and messaging that aligns with the goals of their stakeholders. After the East Palestine Train Derailment in Ohio, the NVFC coordinated with local fire chief Chip

2 | Understanding the History of the Volunteer Fire Service and Fire Department

Figure 2–1. RecruitNY and FASNY volunteer firefighter recruitment poster

Comstock to provide testimony to the Senate Committee on Commerce, Science, and Transportation. The hearing became an opportunity to raise awareness for the need of critical funding for first-responder training, equipment, and health screenings. Chief Comstock also explained how firefighters face a higher risk of developing cancer compared with the general public. He urged Congress to fund baseline and follow-up health screenings for firefighters who have been exposed to hazardous materials. Comstock explained that these screenings would protect firefighters and save resources by preempting potential health complications including cancer.

Chief Comstock's testimony was prepared in advance and reviewed by multiple subject matter experts from across the country to ensure the key priorities and issues within the volunteer fire service were addressed. Thus, the message was kept consistent and coordinated.

What messages are you sending external stakeholders about recruitment and retention issues? Is your message consistent with local and state counterparts? Are you integrating information and data from national groups when you talk about these issues?

Expertise and Research

National fire service organizations often have access to specialized expertise and research on the issues they advocate for. They can provide evidence-based arguments and data to support their positions. For example, National Development and Research Institutes (NDRI)-USA, a private, not-for-profit research and development organization, offers the Center for Fire, Rescue, and EMS Health Research (CFREHR). CFREHR represents a tool "to understand and improve the health of first responders through systematic research and evaluation."[5] NDRI has provided numerous studies showing the impact that health and safety issues have on firefighter retention.

Are you familiar with the national studies and data on recruitment and retention? Do you know how to access current trends and research? Are you aware of what organizations are conducting research and looking for local departments to participate?

Networking and Collaboration

National fire service organizations can facilitate networking and collaboration among different groups and stakeholders. This provides a sense of unity and solidarity. Local fire service leaders who engage with these national groups have the ability to connect with other leaders from across the country who are facing similar issues.

The NVFC Training Summit is in its 10th year, bringing individuals together to discuss and learn best practices for recruitment and retention. The International Association of Fire Chiefs also offers an annual symposium to bring together volunteer and combination department leaders to discuss the needs of the fire service.

Are you connecting with national groups and participating in their annual events? Did you know that they offer scholarships or stipends to attendees? When you attend events such as these, do you network to find other fire service leaders to connect with and share ideas?

Public Awareness and Education

National fire service organizations have the resources to offer nationwide awareness campaigns and educate the public about the issues that their stakeholders are invested in. This generates public support and awareness. For example, the NVFC partnered with John Deere to bring forward the engaging documentary *Odd Hours, No Pay, Cool Hats*, which highlights how everyday community members can be volunteer firefighters.

Did you know that you can offer a showing of this documentary to your community as a recruitment tool? Fire, emergency, and rescue agencies can host a movie screening as a volunteer recruitment event.

Legislative Impact

At the national level, legislation and policies that slip through without fire service stakeholder input can have far-reaching negative consequences. National fire service organizations are better equipped to monitor proposed legislation and policy changes, engage with policymakers, and influence legislative decisions. These actions can have a direct impact on the benefits volunteers view as incentives for volunteering. For example, when I visit Congress.gov and search for "volunteer responder," I get over 2,800 results with the "Legislation" search criteria, and if I search for "firefighter" with the same search criteria, I get over 2,700 results.

Are you aware of what legislation is pending? Do you receive updates on key legislation from your state or a national organization? Do you know how to provide feedback on issues that impact recruitment and retention?

Building a Movement

National fire service organizations can spearhead movements around critical issues and mobilize their membership for widespread support and action. For example, firefighters from both career and volunteer departments have joined together to address occupational cancer. One result of this movement was the development of the National Firefighter Registry for Cancer, which is the "largest effort ever undertaken to understand and reduce risk of cancer among U.S. firefighters."[6]

Addressing Systemic Issues

National fire service organizations can tackle systemic issues that require broader and sustained advocacy efforts. For example, in 2004 representatives of the major fire service interest groups joined together to develop the 16 Firefighter Life Safety Initiatives to inform the safety culture.[7]

Did you know that Firefighter Life Safety Initiative 5 deals with developing and implementing national standards for training, qualifications, and certification (including regular recertification) based on the duties they are expected to perform that are equally applicable to all firefighters? How could such recommendations impact how you train new recruits?

Empowering Historically Marginalized Voices

National fire service organizations can champion the workplace rights and interests of historically marginalized or underrepresented groups, ensuring that their concerns are heard and addressed at the national level. Hearing these voices is critical to develop strategies for recruitment and retention of members of these groups in the fire service.

Dr. Sara Jahnke and colleagues have shown that women, although representing a small proportion of the fire service, are overrepresented as targets of discrimination and harassment.[8] Women in Fire and the NVFC have partnered to develop the Fire Service Discrimination & Harassment Toolkit. This resource aims to help both volunteer and career first responders to prevent, identify, and respond to discrimination, harassment, and retaliation in the workplace.

Are you familiar with resources offered to support underrepresented volunteers? Have you created a workplace environment that fosters communication and inclusiveness for historically marginalized groups? Do you train your existing members on best practices for empowering others?

Examining the History of National Fire Service Organizations

The founding of national fire service interest groups demonstrates the power of stakeholder collaboration. These groups provide an avenue for local volunteer firefighters to engage at a national level. In the next sections, we will look at the history of two such groups that, through their missions, have had a direct impact on how volunteer fire departments operate today.

NVFC

Almost two and a half centuries after Ben Franklin formed the first volunteer fire department, the NVFC was founded in 1976 to serve as a voice for volunteer firefighters at a national level. With representation from 49 state associations, the NVFC is the only national fire service organization dedicated exclusively to supporting, educating, and advocating for volunteer fire and emergency personnel.[9]

The NVFC accomplishes its mission and provides meaningful support to fire service and EMS organizations through a wide range of services and programs:

- Representing the interests of the volunteer fire, emergency medical, and rescue services at the U.S. Congress, federal agencies, and committees that set national standards
- Focusing on health and safety
- Helping departments to recruit and retain fire, emergency, and rescue personnel
- Providing training on topics that matter to you
- Assisting departments in establishing support programs
- Fostering the next generation of firefighters

The NVFC *is* the voice of the volunteer, which includes *every* volunteer firefighter, at the federal level. The positions they take have a direct impact on local fire departments. The NVFC has a seat on NFPA committees and federal work groups and has direct access to members of Congress. Because the NVFC speaks on the behalf of every volunteer fire department in the country, it is important that they have *your* feedback.

Cumberland Valley Volunteer Firefighters Association

Over 70 years prior to the founding of NVFC and well before the beginning of the Society of American Engineers (SAE) and the NFPA, the Cumberland Valley Volunteer Firemen's Association (renamed the Cumberland Valley Volunteer Firefighters Association in 2023) (CVVFA) was founded in 1901 to establish the hose thread size to be used by volunteer fire departments to protect railroads of all towns in the valley between Harrisburg, PA, and Winchester, VA.

After losing two members in struck-by incidents, CVVFA established the national Emergency Responder Safety Institute (ERSI) in 1998 to

> serve as an informal advisory panel of public safety leaders committed to reducing deaths and injuries to America's Emergency Responders. Members of the Institute, all highly influential and experts in their fields, are personally dedicated to the safety of the men and women who respond to emergencies on our nation's streets, roads, and highways. Members of the Institute include trainers, writers, managers, government officials, technical experts, and leaders in the public safety world who, through their individual efforts, and collective influence, can bring meaningful change.[10]

Since the implementation of the ERSI, CVVFA has had an impact on reducing injury and fatalities on the nation's roadways. CVVFA has been awarded many federal grants and contracts to tackle this issue at the national level. This has been done through advocacy, educating over 100,000 first responders in person or online (at ResponderSafety.com), lobbying for legislative changes, and having a voice at the national level. CVVFA is an example of how leveraging stakeholders can be a powerful tool in creating social change.

How does CVVFA impact you locally? Have you ever wondered why we are required to wear high-visibility vests while working the roadway? Did you ever wonder where the chevron striping on apparatus came from? Do you know who developed the curriculum for the National Traffic Incident Management Responder Training certification? CVVFA's voice was partly responsible for each of those things.

Other National Fire Service Organizations

The NVFC and CVVFA are just two examples of how the past has paved the road to the future for the volunteer fire service. The reach of these two organizations demonstrates why fire service leaders should engage with national organizations. Other organizations with focus on volunteer fire departments include the following:

- *International Association of Fire Chiefs—Volunteer & Combination Officers Section.* Mission: To develop and enhance effective, professional leaders of the volunteer and combination fire service by providing tools, resources, and representation to lead their organizations effectively.
- *Women in Fire.* Mission: Leading the fire service community by providing training, education, advocacy, resources, mentoring, and networking to enhance the fire service.
- *The Science Alliance.* Science to the Station: A Health & Wellness Alliance is a community of scientists, health care workers, and fire service professionals dedicated to providing critical information needed to improve the safety and health of firefighters and other first responders.
- *CFSI.* Mission: To educate members of Congress about the needs and challenges faced by our nation's fire and emergency services, to help legislators to understand how the federal government can support the needs of local first responders.

STRATEGY STOP

Educate your volunteers on the history of your volunteer fire department.
Understanding how the fire department was founded allows us to appreciate the dedication of past volunteers, learn from their experiences, and strengthen future endeavors. It also highlights the enduring value of volunteerism and its continued importance in contemporary firefighting efforts.

Developing methods to help new volunteers to understand how the volunteer fire department started is important for several reasons:

- *Historical context.* Knowing the origins of the local fire department can provide new volunteers with the historical context and help them to appreciate the evolution of the department and the significant contributions made by volunteers throughout the fire department's history.
- *Legacy and tradition.* The traditions and values established by early volunteer firefighters often shape the culture and identity of today's fire department. Reflecting on the legacy of past members can foster a sense of pride and camaraderie among today's volunteers.
- *Appreciation for volunteerism.* Learning about the earlier years of volunteer first responders within the community emphasizes the spirit of community service. This appreciation can inspire present and future generations to volunteer with their local fire department.
- *Lessons learned.* Studying the challenges and triumphs of early volunteers can provide valuable lessons for shaping today's fire department. Understanding how past volunteer departments overcame limited resources and support can offer insights for addressing current issues.
- *Innovation and progress.* The history of the volunteer fire service can showcase the progress made in tactics, equipment, and the department in general over time. Awareness of past challenges and solutions helps us to recognize the advancements that have improved fire safety and prevention and that encourage future change.
- *Connection to the community.* Volunteer fire services often have deep roots in their communities. In many communities, the fire department has been as influential as the local churches in shaping beliefs and bringing people together. Understanding the influence of past connections allows present-day volunteers to develop methods to foster stronger connections with local community members and understand the community's needs better.
- *Preservation of knowledge.* The history of your volunteer fire department may include unique institutional knowledge. Preserving this knowledge is important to ensure that these techniques and practices are passed down through generations.
- *Inspiration for change.* Studying the evolution of the volunteer fire service can inspire improvements in current practices and encourage innovation to tackle modern challenges effectively.

STRATEGY STOP

Join a national fire service organization and get involved to amplify your voice.
Local volunteer fire service leaders can find several benefits in joining national organizations. Participating in such groups can lead to several benefits:

- *Networking opportunities.* National groups provide a platform for leaders to connect and network with their counterparts from various regions. This allows them to share experiences, best practices, and valuable insights that can enhance their leadership skills and improve local firefighting efforts.
- *Access to resources.* National organizations often offer access to a wealth of resources, including training materials, research findings, and up-to-date information on firefighting techniques, equipment, and safety protocols. This can be especially valuable for local leaders seeking to stay informed and implement the latest advancements in their communities.
- *Training and professional development.* Membership in national groups typically comes with opportunities for training and professional development. Leaders can attend conferences, workshops, and seminars to enhance their knowledge and skills, ultimately benefiting their local fire service.
- *Advocacy and representation.* National organizations serve as powerful advocates for firefighting interests at a broader level. Joining such groups allows local leaders to contribute to discussions on policies, regulations, and funding that can have a significant impact on their communities.
- *Collaborative initiatives.* National groups facilitate collaboration among local fire service agencies. By working together on initiatives, leaders can address common challenges, pool resources, and implement more effective strategies for fire prevention, emergency response, and community safety.
- *Standardization and accreditation.* National organizations often play a role in setting standards and establishing accreditation programs for fire service agencies. Local leaders who participate in these programs can ensure that their departments adhere to the best practices and meet recognized standards.
- *Information exchange.* Being part of a national group allows leaders to stay informed about trends, research findings, and emerging issues in the firefighting community. This exchange of information can help local leaders to adapt and respond effectively to evolving challenges.

- *Professional recognition.* Membership in national organizations can enhance the professional standing of local fire service leaders. It demonstrates a commitment to ongoing learning, collaboration, and staying connected to the broader firefighting community.
- *Grant and funding opportunities.* National groups may offer information about—and even facilitate access to—grants and other funding opportunities. This can be particularly helpful for local fire services looking to improve infrastructure, acquire new equipment, or implement community education programs.
- *Crisis response and mutual aid.* National organizations can facilitate mutual aid agreements and coordination during large-scale emergencies. This ensures that local fire services can effectively respond to crises with support from neighboring or national resources.

Thus, joining national groups can provide local volunteer fire service leaders with a range of benefits—from improved networking, to access to resources, and to opportunities for professional development and advocacy on behalf of their communities.

Review Questions

1. What does history provide fire service leaders with?

2. The volunteer fire service was started in _____, when _____ founded the Union Fire Company in Philadelphia.

3. The NVFC was founded in 1976 to serve as what?

Discussion Questions

1. What is the history of your fire department?

2. What methods can your fire department use to educate new firefighters on the history of your fire department?

3. How can your department leverage national organizations, and what organizations do you belong to?

3

THE VALUE OF THE VOLUNTEER

Value your volunteers. Treat the newest member as a future chief of the department.

—Stephen Marsar, battalion chief
Fire Department of New York (FDNY)

Common Misconceptions about Volunteer Firefighters

Misconceptions about volunteer firefighters are not uncommon among the public and within the fire service industry. Being aware of common misconceptions and knowing how to respond are important to fire service leaders working to recruit new members. Negative stereotypes can influence an individual's decision to serve. Addressing misconceptions about volunteer firefighters will foster a better understanding and appreciation for the vital role that volunteer firefighters play in protecting and serving their communities. Thus, debunking misconceptions proactively can have a direct, positive impact on recruitment activities.

The following are common misconceptions about volunteer firefighters that fire service leaders will need to provide clarity to external stakeholders:

- *Lack of training.* Many assume that volunteer firefighters receive less training than their career counterparts. This leads to doubt regarding skills and ability. In reality, the majority of volunteer firefighters undergo the same training at the same academies as do their career counterparts. Volunteers receive extensive training and certifications to ensure that they are well-prepared for addressing emergencies within their community. How do you educate the public that your volunteer firefighters are adequately trained? Do you share on social media—for example, when a volunteer gets a new certification and pictures or video of your firefighters actively training?

- *Volunteers are less skilled or committed.* There's a misconception that volunteer firefighters are less skilled or less committed than career firefighters because "they don't do it as a job." In reality, most volunteer firefighters have the same level of expertise and dedication. They are often called on by neighboring career fire departments to provide mutual aid during large incidents. These volunteers not only respond to fires and EMS calls but also serve as community educators informing the public about preparedness and fire and life safety. Fire chiefs of volunteer departments deal with budgets and must navigate the political system in the same way career chiefs do. How do you let the public know that your firefighters are skilled? Do you share statistics on social media, such as your call volume and mutual aid assists?
- *Volunteering in the fire service is only for younger generations.* While some may believe that volunteer firefighting is predominantly for younger generations, many fire departments have and recruit volunteers of various ages. There is a job for everyone to serve within the volunteer fire service. Of volunteers who return to a second year of service, 66% are primarily from either the baby boomer or mature generation age group. The volunteer fire service can fill the need for purpose and a sense of belonging as this generation retires. How do you communicate to the public that there is a place for all, despite age? Do you share on social media photos of members in action from different generations?
- *Volunteers are needed to fight fires only.* Volunteer firefighters are called on for many emergencies, not just fires. They respond to medical calls, traffic accidents, hazardous material incidents, wellness checks, and other emergencies within their community. Nonemergency volunteers are just as valuable as those on the front line fighting fires. Does your department provide volunteer opportunities for nonemergency operations? Is the public aware that you need webmasters, mechanics, accountants, recruitment coordinators, educators, rehabilitation specialists, and fundraiser support, among other nonemergency roles?
- *Lack of professionalism.* Some may assume that volunteer firefighters lack professionalism or dedication compared to career firefighters. However, volunteers often approach their duties with the same

level of commitment and pride in their work as their career counterparts. In fact, the public demands professionalism, and often community members do not even realize the people responding to their emergency are not paid. Does your department operate as a professional business with requirements and expectations?

- *Volunteers are less equipped.* While volunteer departments may have budget constraints, they are often equipped with modern firefighting equipment to perform their duties safely and effectively. The lack of a payroll allows financial resources to be earmarked for equipment. Is the public aware of the equipment that you own to serve them in their time of need? Do you share on social media photos or video of volunteer firefighters learning to use new equipment or performing drills with existing equipment?
- *Volunteering is a short-term commitment.* There's a misconception that volunteer firefighting is a short-term commitment or merely a stepping-stone to a career. Many who transition into career firefighting still volunteer within their communities or become advocates for using volunteer opportunities as a career path. Is your organization willing to be flexible with requirements to allow individuals to serve long-term? Are you allowing for quality engagement instead of mandating quantity?
- *Only men can be volunteers.* Historically, firefighting was a male-dominated field, leading to the misconception that only men can be volunteer firefighters. The term "fireman" also did not convey an inclusive message. While the percentage of females in the fire service are low, numerous female volunteers excel in the field (fig. 3–1). How do you inform girls and women in your community that there is an opportunity for them to volunteer within your organization?
- *Volunteers don't respond quickly.* Some people assume that volunteer firefighters might take longer to respond to emergencies because of their off-duty status. However, many volunteer departments have efficient response systems and are able to meet the National Fire Protection Association (NFPA) response requirements. Do you share with your community that your volunteers are actively responding as soon as a 9-1-1 call comes in?
- *Misconception of time commitment.* Another set of misconceptions among the public is that being a volunteer firefighter means just showing up in your spare time and that the only duties volunteers within the fire service have are fire suppression and emergency medical services (EMS) response. Community members often are

Figure 3–1. Dr. Candice McDonald being sworn in as the first female president of the Cumberland Valley Volunteer Firefighters Association

unaware of the level of commitment it takes to become and remain an active volunteer. Moreover, volunteers also play key roles in running fire service organizations, by overseeing daily operations and generating revenue for equipment. Without volunteers stepping up to perform these tasks, many fire service organizations could not exist. Even as some fire service organizations expand and hire paid staff, they still rely on volunteers to carry out work necessary for operations. Volunteers are critical to the survival of volunteer fire service organizations. Is your community aware of all of the time your volunteers commit to provide them with fire and life safety coverage?

Importantly, the fire service is not the only organization that relies heavily on volunteers. In fact, in 2000 volunteers in the United States worked the equivalent of more than 9.1 million full-time positions, with a combined savings of $239 billion.[1] These numbers show a need for service among various organizations and a desire to serve among volunteers.

What misconceptions does your community have about the fire service and about your department? How have you worked to address misconceptions? Are you actively sharing with the public your activities and successes?

Focus on the Value of the Volunteer

We must shift to a positive mindset when we talk about volunteerism within the fire service. After all, why would anyone want to join an organization headed for devastation?! Some fire service leaders have openly stated that they feel like the volunteer fire service is destined for extinction and that shifting to staffed stations is the only solution. However, the data tells us that volunteerism in America is not dead. Many citizens are civic minded and want to serve their communities; the problem is that many fire departments have been using the same playbook for recruitment and retention operations for decades. It is time to rewrite the playbook and create an environment that shows potential volunteers what is in it for them and sell them on why they should join.

The Value of Volunteering to the Community and to the Volunteer

Before we can sell the concept of supporting volunteer firefighters to others, we need to understand the value that volunteer fire service and EMS personnel bring to the communities they serve and to the individuals engaging as volunteers. Information about the value of volunteer time can be persuasive during conversations with financial stakeholders. Being able to communicate with factual data about the cost savings of volunteer firefighters to a community may lead to an increase in community support, and this information can be shared strategically, as leverage in grant proposals and in annual reports.

Additional benefits of community volunteers in the fire service and EMS include the following:

- With a greater diversity among first responders comes a higher chance of reflecting the community being served. We can't assume that a practice that works in one community will work in another. One size does not fit all, and best practices do not always translate across cultures. Certain practices can easily be adapted, while others can't be implemented within a certain geographic or demographic area. Does your organization look like the community you serve?
- Diverse occupations of volunteer members means an increased skill set among fire service personnel. Use of an individual's skill set is one motivating factor for volunteering. In fact, one study showed that 75% of volunteers volunteered to apply their skills and experiences to help others.[2] Have you mined the skill sets of those within your community to meet your organizational needs?
- Volunteers comprise a vital part of emergency response management. The American Red Cross has reported that volunteers respond to 65,000 disasters a year.[3] In 2021 fire departments responded to over 36.5 million calls.[4] Is your community aware of the number of calls your volunteers respond to each year and how that benefits them?

Volunteers expand the reach of the community toward support and resources. Volunteers bring not only valuable time and energy but also avenues to diverse resources. Volunteers can advocate for the organization and increase community support within their circles of influence. Are you aware of the different groups your volunteers are engaged with outside of the fire department? Do you know what resources they have access to, such as equipment or talent?

Volunteers make up 65% of all U.S. fire services and donate approximately $46.9 billion per year worth of time each year[5] (fig. 3–2). The *cost value of volunteer time* is estimated based on the average earnings of private-sector employees as reported by the U.S. Bureau of Labor Statistics and the annual average hourly earnings an organization would have to pay should they have to hire paid workers to fulfill these roles. Importantly, the value of a volunteer is *not* based on the amount of money a volunteer earns within their paid occupation; it is solely based on the services they provide in their volunteer profession. In addition to the hourly wage amount, an additional fringe-benefit rate cost should be added. The fringe-benefit rate used by the Independent Sector estimates an hour of volunteer time at 15.7%.[6]

Fire service leaders can calculate the value of volunteers within their organization by adding all the hours of time spent per member (responding to calls, training, and conducting fire department operations during nonemergencies) and multiplying that number by the pay rate listed in the U.S. Bureau of Labor

Figure 3–2. Infographic showing the monetary value of time donated by volunteer firefighters in the United States

Statistics Occupational Outlook Handbook. Once that is calculated, the 15.7% fringe-benefit rate can be added to give the overall cost value of volunteer time in your department.

Do you know the cost value of your members? Does your community know the cost value of the time your volunteers donate each year? Do your volunteers know how valuable their contribution is to the community?

STRATEGY STOP

Learn to calculate volunteer firefighter time contributions to leverage external stakeholder support. Fire service leaders can use the volunteer firefighter time value calculation demonstrated here to determine the value of the time their volunteers donate. This figure can be used to educate external stakeholders on the value of the time volunteer firefighters contribute, to leverage resources for support of operations.

Volunteer Firefighter Time Value Calculation
There are several sources to calculate the time value of volunteer firefighters.

U.S. Bureau of Labor Statistics Data
- The U.S. Bureau of Labor Statistics Occupational Outlook Handbook median annual wage for firefighters and prevention workers (May 2021) is $50,930[7]
- 2,080 hours a year is equivalent to one person working 40 hours per week for 52 weeks
- Divide the median annual wage of $50,930 by the 2,080 hours per year worked: $50,930 ÷ (2,080 hours) = $24.49 per hour
- Add the fringe-benefit rate to the hourly rate: $24.49 × (100% + 15.7%) = $28.33, which is the hourly value of a volunteer firefighter or prevention worker
- Overall value of a volunteer firefighter's time = $28.33 × (no. of hours a member donated)

The Independent Sector Data
- The Independent Sector sets the hourly value of volunteers at $33.49[8]
- Overall value of a volunteer firefighter's time = $31.80 × (no. of hours a member donated)

Plan to Track Volunteer Data

Develop a plan for tracking fire service volunteer data to aid your organization in determining the value of the service provided by your volunteers. Volunteer hours also tell a story of what resources are needed to accomplish the fire department's mission. Collected volunteer data can guide future decisions of fire service leaders for improvement and expansion of recruitment and retention activities. Volunteer hour tracking can also have the following tangible impacts on the fire department:

- Data provided on volunteer hours can be used as in-kind donations for securing grants to support the fire department.
- Volunteers can use a log of hours to leverage scholarships, meet certain educational requirements, or qualify for special awards.
- For volunteers who are employed by organizations that are active in corporate philanthropy, keeping a log of hours could lead to a donation of funds from their employer based on volunteer time contributions with the fire department.

The best plan for tracking your volunteer data will depend on the specific needs and resources of your fire department. Remember that the success of a volunteer data–tracking plan requires the commitment and cooperation of all stakeholders involved. By implementing an efficient and organized system for tracking data, fire service leaders can better support volunteer needs, improve engagement efforts, and enhance the overall effectiveness of the department. The following list gives strategies for creating an effective plan for tracking volunteer data:

- *Assign a volunteer manager.* A volunteer manager can be appointed to keep records of all volunteer data. This person would be responsible for collecting all of the information that will allow you to gain a true understanding of the value and cost of your volunteers. Can you recruit a nonemergency volunteer to serve in this role?
- *Identify data objectives.* Clearly define the objectives that your fire department has for tracking volunteer data. Determine what information you need to collect and how it will be used to shape future planning. Are you familiar with the data you need for grants, staffing, and other operational activities?
- *Choose a data management system.* Select a data management system that suits your fire department's size and requirements. It could be as simple as a spreadsheet or as extensive as specialized volunteer

data management software. Do you have funding to purchase data management software, or is there someone within your community who is fluent in Excel and willing to design a data-tracking spreadsheet for you?

- *Define needed data categories.* Create categories for different types of volunteer data, such as personal information, demographics, contact details, training, and volunteer hours for emergency and nonemergency tasks performed. What categories are important to your organization and community?
- *Implement consistent data entry procedures.* Establish guidelines for consistent volunteer data entry, to ensure accuracy and uniformity of the information recorded. Determine who will be responsible for data entry and how often they will input data. Will you expect a monthly report to share with leadership and stakeholders?
- *Train volunteers on data usage.* Provide training to volunteers responsible for data entry. Emphasize the importance of accurate, uniform, and timely data inputs. Provide training to the department on the importance of data. Explain its value and how it will be collected. Is there someone in your organization or community who can train your volunteers?
- *Use secure data storage.* Information on potential and current volunteers often contain *personal identifiable information* (PII). Ensure that all volunteer data is stored securely and in compliance with data protection regulations to protect volunteers' privacy. PII is defined by the U.S. Department of Labor as follows: "Any representation of information that permits the identity of an individual to whom the information applies to be reasonably inferred by either direct or indirect means. Further, PII is defined as information: (i) that directly identifies an individual (e.g., name, address, social security number or other identifying number or code, telephone number, email address, etc.) or (ii) by which an agency intends to identify specific individuals in conjunction with other data elements, i.e., indirect identification. (These data elements may include a combination of gender, race, birth date, geographic indicator, and other descriptors). Additionally, information permitting the physical or online contacting of a specific individual is the same as personally identifiable information. This information can be maintained in either paper, electronic or other media."[9] How will you ensure that your department's data is protected?
- *Provide regular updates.* Regularly evaluate and update volunteer data to keep the collected content limited to what is current

and relevant. Encourage volunteers to inform the assigned volunteer manager about any changes in their contact information or certifications. What reporting requirements will you implement? How will you communicate those reporting requirements to your members?

- *Monitor engagement metrics.* Track metrics related to volunteer engagement, such as the number of volunteers, retention rates, hours contributed, and areas of involvement. Who will be responsible for looking at trends? How will that information be used and communicated?
- *Use reports and analytics.* Use data analysis tools to generate reports and analytics. These can provide valuable insights into recruitment and retention goals, identify areas needing improvement, and help evaluate the overall program effectiveness. What benchmarks, or comparable points of reference, can be used to assess success of your organization's efforts?
- *Feedback and evaluation.* Gather volunteer feedback, and use that data to understand their experiences and improve engagement and satisfaction. This can be done through qualitative data (interviews, open-ended questions, etc.) and quantitative data. How will you improve your data-tracking system? What questions will you ask, and how often will you survey your stakeholders?
- *Continuous improvement.* Regularly review and refine your volunteer data-tracking plan based on feedback and changing needs. Stay open to incorporating new technologies and methods for better data management. How often will you examine strategies for data-tracking best practices?

STRATEGY STOP

Use an app for volunteer management tracking to record how many hours your volunteers donate to the organization. One example is Point (fig. 3–3), a mobile application that is free for nonprofit organizations. In Point, volunteers can input their hours, those hours can then be reviewed and approved by the organization, and the organization can run a report of volunteer activity. The app's dashboard can show how many hours were volunteered and the estimated economic value. The Point administrator can change the labor cost per hour in the settings to match industry standards. Point also can also be used to post opportunities and shifts volunteers can sign up for.

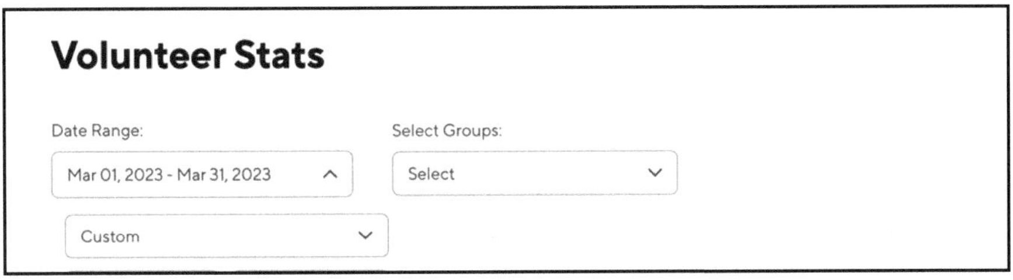

Volunteer Stats

Date Range: Mar 01, 2023 - Mar 31, 2023 — Custom

Select Groups: Select

Figure 3-3. Screenshot of tracking report on the Point app

STRATEGY STOP

Create an Excel spreadsheet for volunteer management tracking of hours, demographics, skills, gaps, value, and trends. Excel is a basic Microsoft Office tool that works as spreadsheet editor. Excel features calculation and computation capabilities, graphing tools, pivot tables, and a macro programming language. The PivotTable functionality in Excel allows you to calculate, summarize, and analyze data to make comparisons and detect patterns and trends. Microsoft Support provides step-by-step videos for creating a pivot table to analyze data. (You can visit support at https://support.microsoft.com and type "create a PivotTable" to get a full list of help articles and videos.) Does your local high school or community college have business students needing a project for their academic requirements who you could recruit to help build your Excel charts?

STRATEGY STOP

Use volunteer data to support department efforts.
Data on the value of volunteer time can be used in various ways to benefit the fire department and community. By knowing the value of volunteer time, fire departments can not only showcase the importance of volunteer contributions but also attract more support, resources, and recognition. The following are ways to apply data on the value of volunteer time:

- *Grant applications.* When applying for grants or funding, fire departments can include the value of volunteer time as an in-kind contribution, demonstrating the additional resources and support the

organization receives. Many grants have a short turnaround period from the time of announcement to the application deadline. Do you have the data needed for funding opportunities easily accessible?
- *Public relations and marketing.* Highlight the value of volunteer time in public relations and marketing materials to showcase the impact and dedication of volunteers, which can lead to more support and recognition from external stakeholders. How are you sharing your data to increase public support?
- *Bolstering reports.* Include the value of volunteer time in reports to quantify the fire department's overall contributions to the community and draw attention to the significance of volunteer efforts. What data are you sharing with funders, elected officials, and other entities to increase support?
- *Fundraising efforts.* Incorporate the value of volunteer time in fundraising asks to show potential donors the added value and efficiency of the fire department's work. Many potential vendors are unaware of the high value that volunteer firefighters bring to their communities. Are you sharing the value of your volunteers when you are talking to potential donors?
- *Board, trustee, and city council reports.* Sharing the value of volunteer time with the fire department's oversight board helps them to understand the resource contributions from volunteers and their importance in achieving organizational goals. Are you providing data charts in your written reports and presentations?
- *Resource allocation.* Understanding the value of volunteer time can help fire departments allocate resources more effectively and determine resource needs. How does data determine your resource allocation?
- *Policy advocacy.* Use the value of volunteer time when advocating for policies and legislative issues. How can you use data to support policy?
- *Volunteer recognition.* Acknowledge the valuable contributions of individual volunteers by highlighting the collective value of their time and effort in written and verbal communication. Do you have an accurate picture of each volunteer's contributions with training, calls, and other support functions?
- *Community engagement.* Share the value of volunteer time with the community to foster awareness of the fire department's impact. This not only shows the value of the volunteer but could also encourage others to volunteer with the fire department. How do you share the value of your volunteers' contributions with the public?

- *Collaboration opportunities.* Demonstrating the value of volunteer time can open doors for collaborations with other organizations and businesses. When you look at your data, is there a gap or need trend that could lead to a collaboration with a community partner to address it?

STRATEGY STOP

"Play-pay" your volunteers quarterly based on the data collected.

One method to recognize the value volunteers make is to issue a *play-pay* certificate or picture highlighting each volunteer's contributions each quarter (fig. 3–4). This can be easily done with frequently used software tools (e.g., Microsoft Word or Canva). Turning volunteer hours into visual data lets volunteers see the difference that they are making and can act as a motivator to keep donating. Organizations that express their appreciation for their volunteers see an increase in volunteer satisfaction and willingness to volunteer further. Some volunteers can turn in donated hours to their employer for matching contributions. Have you asked your volunteers if their employers have a matching or other donation program?

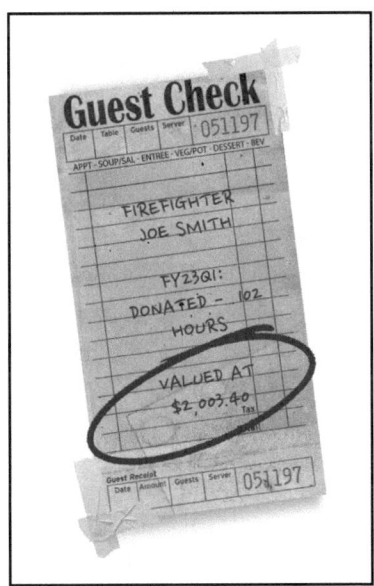

Figure 3–4. Example play-pay certificate

Communicating the Value of *Your* Volunteers

> *Career or volunteer firefighters face largely the same risks on the job. The IAFF, NVFC, and the IAFC often work side-by-side to identify and influence legislation and standards that positively affect the health and safety of all first responders. Career and volunteer firefighters can only benefit by working together with one voice towards a common goal of creating a safe, healthy working environment.*
>
> —Sarah Lee, chief executive officer
> National Volunteer Fire Council (NVFC)

It is important for fire service leaders to communicate the value of their volunteers in a way that bridges stakeholder relationships, rather than in a way that causes division. While volunteer firefighters should be prideful of the cost savings they bring to their localities, they must not discount or speak negatively about the cost of career firefighters. Career firefighters not only offer a valuable service to the communities they serve but also influence and are advocates for industry safety.

The International Association of Fire Fighters (IAFF) was formed in 1918. The IAFF has worked alongside other groups, such as the NVFC, to promote legislation to support and protect firefighters and their families through positive changes to benefits, presumptive laws, and personal protective equipment. The IAFF and its members should not be viewed as a threat but as allies working to improve firefighter health and safety.

How a fire service organization communicates the value volunteers bring to the community matters, to avoid opposition from members of the IAFF and the general community. For example, running a print advertisement like the following could trigger mixed emotions: "Last year, Any Town Fire Department volunteers donated X number of hours to serving the community, the equivalent of Y paid firefighters." Some might interpret that message as "Any Town Fire Department kept Z people from being employed" or "Why does Any Town Fire Department need so much money when they operate on all volunteers?" The positive outcome from those donated hours is missed when we promote hours donated over the success stories that we should be conveying. Additional negative consequences of focusing solely on replacement value (volunteer firefighter instead of a paid firefighter) include the following:

- The misconception that volunteer firefighters are 100% free, when in fact they come with costs (time, training, gear, insurance, etc.)

- The belief that the number of hours donated by volunteers is how success is determined, when in fact success should be based on customer satisfaction and community impact
- The attitude of the volunteer that the organization owes them because they are donating a value amount instead of a gift of service

Do the messages your department communicates internally and externally strengthen relationships? Have you communicated with your volunteers the need to not cause division with career brothers and sisters? Have you met with your partner career departments to form collaborations?

STRATEGY STOP

Create a fire department fact sheet of what your volunteers have accomplished to date, in the past month, and over the past year. Fact sheets can provide fire service leaders with a document to hand to stakeholders and journalists with key information. Include the following information:

- How many fire calls did they respond to?
- How many EMS calls do they run?
- What was the top type of incident?
- How many fire prevention activities did they engage in?
- What projects are they currently working on?
- What projects did they finish?
- What are the demographics of the community they serve?
- What is the total value of time that members as a whole donated?

Tips for Developing a Fire Department Fact Sheet

- Limit the fact sheet to one page in length.
- Keep information brief and concise.
- Include visually appealing graphs, photos, charts, and bullet points.
- Ensure that facts are easy to read and understandable to those outside of the organization (fig. 3–5).
- Include a call to action on every fact sheet with a form of contact information (website, social media links, email, phone, etc.).
- Use tools like Canva to help create an appealing fact sheet.
- Save the fact sheet in multiple formats (e.g., as PDF and JPEG files), for different sharing options.
- Be sure all members are familiar with and have access to the fact sheet, for consistent messaging.

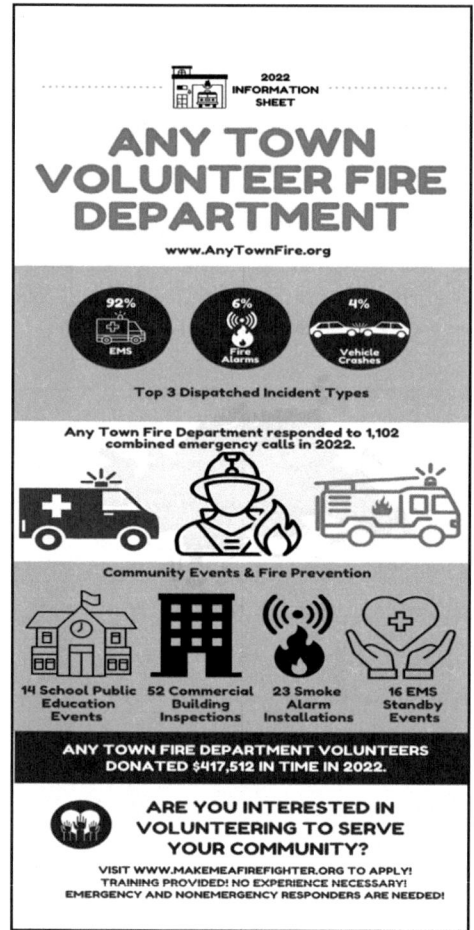

Figure 3-5. Example fire department fact sheet

STRATEGY STOP

Incorporate national volunteerism data to share a story of the value of volunteers in general.
AmeriCorps offers a toolkit with already created graphics for each state. There is no charge or permission needed to use this toolkit (available online at https://socialpresskit.com/americorpstoolkits#volunteering-and-civic-life-in-america). Share the graphics with a social media post that includes a call to action for your local department (fig. 3–6).

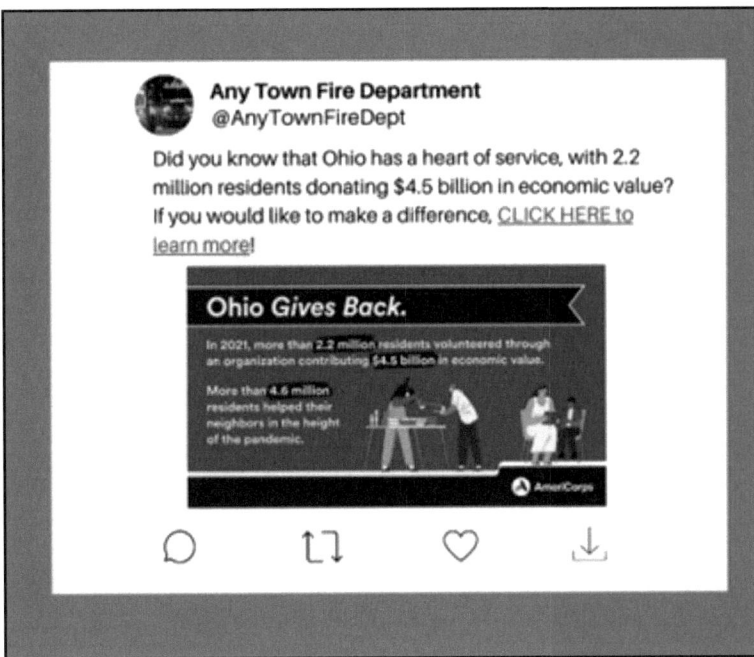

Figure 3-6. Example social media post using premade graphics from AmeriCorps

STRATEGY STOP

Showcase success stories shared by community members.
Often fire service organizations receive thank you letters and cards, social media messages, and family drop-ins giving thanks for a job well done. Ask the individuals expressing gratitude if it is okay to share their message on social media or with stakeholders. Some families may even agree to be photographed or create a short video expressing their thanks that you can use to showcase department success stories. It is good practice to have those individuals sign a release. You can also share the stories of your firefighters participating in feel-good community moments, such as teaching firefighter safety to children (fig. 3–7).

Tips for Sharing Community Member Success Stories
- Do ask for permission before sharing messages or photos expressing community member gratitude.

Figure 3–7. Example social media post showcasing community engagement

- Do ask community members to pose in front of department apparatus, with the department name showing, for photos and videos that will be used on social media.
- Do ask the community members for permission to share their name or any other identifying information.
- Do review the content of the social media post with the community member before sharing.
- Do develop different types of visuals to share across diverse platforms.
- Do let community members know they are welcome to share social media posts.
- Do have any non–fire department members sign a photo release authorizing you to use the photos taken (fig. 3–8). Sample photo releases can be found by searching the Internet for "government sample photo release" (one such release is available from the National Institutes of Health to download at no cost from https://www.nih.gov/sites/default/files/institutes/ocpl/sample-photo-release.pdf).

***Note: This is a sample release. You will need to identify appropriate language for your circumstances.**

NIH National Institutes of Health
Turning Discovery Into Health

National Institutes of Health
Office of Communications
1 Center Drive, MSC 0188
Bethesda, MD 20892-0188
(301) 496-4461
FAX (301) 496-0017

AUTHORIZATION

I, ______________________, authorize The National Institutes of Health (NIH) to record and/or broadcast interviews, films, recordings, or photographs of me taken with my knowledge and agreement. The recordings may be used by NIH for the development, promotion and broadcast or distribution in any medium of science, health or educational programming. NIH is entitled to edit, copy, adapt or translate the contribution and authorize others to do so in connection with NIH projects.

SAMPLE

Date ______________________ Printed Name ______________________

Signature ______________________

Email address: ______________________

Figure 3–8. Sample photo release

Review Questions

1. Name at least three common misconceptions about volunteer firefighters.
2. What two steps are involved in identifying objectives to create an effective volunteer data–tracking plan?

Discussion Questions

1. How is your department tracking volunteer value?
2. In what ways does your department communicate the value of your volunteers?

4

FIND OPPORTUNITIES TO EXPRESS VOLUNTEER VALUE

Of all first responders, perhaps Firefighters have traditionally been best at outreach, sharing their story and mission with children and the community through its organizations and institutions. In turn, Firefighters have the best public relations and are a model from which all others can learn.

—Bruce Williams, past president, Salem, Ohio Kiwanis

Identify Community Engagement Opportunities

Engaging in existing community events can bring great value to your organization. Identify opportunities for your department to host and attend along with community members to engage with them and educate them about what your fire service organization does and the impact that your volunteers have on your community. These types of events can also be used as a call to action for fundraising and recruitment (fig. 4–1).

What existing community events can you partner with? Is there a certain annual event that brings people out into the community when you could offer an open house? Have you had conversations with other community leaders about ways to collaborate and gain exposure?

STRATEGY STOP

Host a listening tour with an intentional guest list of community leaders, influencers, and stakeholders you serve.
This event can be an opportunity to share your mission and the work you have done over the past year. In addition, it should serve as a platform for

Figure 4–1. Winona (OH) Fire Department engaging with youth at a community event at Woolf Farms

your constituents to provide input and insight on what gaps exist and on how you can recruit new volunteers, as well as hearing from individuals you would not otherwise. It will be important to follow up with community leaders after the event with opportunities for them to get involved with the fire department or for the fire department to engage with them.

Who in your community do you need to connect with? Are there geographic areas where you need to strengthen your relationships with the community, and if so, who can help you access those areas? Are there cultures within your community whose needs you require help to understand, and if so, what leaders can help you reach them?

STRATEGY STOP

Make an outreach plan that includes existing social groups within your community.
Reach out to church groups, school groups, parent groups, and civic service groups such as Kiwanis, Rotary, Ruritans, and Lions Clubs. Remember to offer to meet a need of the group you are reaching out to; good marketing is about meeting the needs of stakeholders.

Who could you include in a brainstorming session to develop an outreach plan? Are there people outside of your organization who could provide input?

4 | Find Opportunities to Express Volunteer Value

STRATEGY STOP

Develop canned email and phone scripts to use for outreach to community groups for requesting to make a presentation to their group.

Sample Email Request to Make a Presentation

Dear Mr. Williams:

The Any Town Fire Department is reaching out to different civic groups within the community to strengthen our partnerships. We would love the opportunity to attend one of your meetings to share what is new at Any Town Fire Department and to share a life safety message with your members. We are also very interested in learning more about your organization and identifying ways we can aid your organization. I have also attached to this email Any Town Fire Department's 2022 Fact Sheet. Please feel free to call or email me with questions or to set a date for a presentation.

Sincerely,
Firefighter Jane Anytown
Outreach Coordinator, Any Town Fire Department

Sample Phone Request to Make a Presentation
Hello, Mr. Williams. My name is Lt. Joe Brown with the Any Town Fire Department, and we are interested in meeting with members of you organization to find ways we can strengthen our partnership. We would love the opportunity to attend one of your meetings to share what is new at Any Town Fire Department and to share a life safety message we think your members will find valuable. Would you be interested in getting a date and time on the schedule?

STRATEGY STOP

Have members create an outreach lead tree to identify community outreach opportunities for you to present about your fire department's volunteer needs. A lead tree starts with the people you know and builds out along the connections that come with those relationships.

The following are some brainstorming questions to ask yourself as you fill in the boxes of your lead tree (fig. 4–2).

Are you connected with a church?

- What small groups within the church would be open to hearing a presentation?
- What already-occurring events at the church could you ask to have an information table at (Celebrate Recovery Vacation Bible School, Potluck Sunday, etc.)?
- Does your church have a mission team with members with a heart of service?

Do you have connections with any civic groups?

- Service organizations (such as Kiwanis, Rotary, Ruritans, and Lions Clubs) are always looking for guest speakers. How can you help to fill this need?
- When do your local small business associations meet? Not only can you introduce the members to the fire department, but you can also open the floor to answer questions about fire inspections.

What about the people in your circle of influence?

- Do you have a neighbor or friend in a school parent group who could help you to get a table or spot presenting at the next parent-teacher association meeting?
- Do you have a neighbor or friend who works at a large manufacturing company who could arrange for you to speak at a company meeting or schedule to have a table in a break room?
- Do you have a neighbor or friend who coaches a youth sport or plays in a recreational league? Ask if you could be on first-aid standby with a recruitment table at the games.

STRATEGY STOP

Create a canned presentation that fire department members can use to showcase the value the fire department brings to the community.

A canned presentation is a standard presentation that an organization uses to present information systematically and consistently. This presentation should also talk about the organization's mission, successes, and opportunities for engagement. Creating a canned presentation ensures that no matter who is engaging with the public, the general message stays consistent.

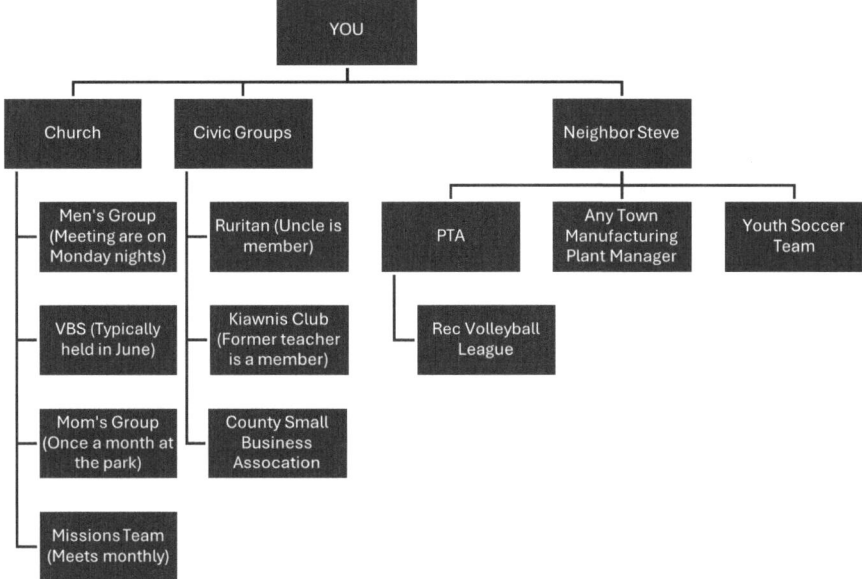

Figure 4-2. Sample outreach lead tree

Slide Content Should Follow This Sequence

Slide 1: Fire department's name and an attention-grabbing photo of volunteers in action. Don't be afraid to stage a photo; this ensures that all safety measures are in place and the picture is high quality.

Slide 2: Tell the story of Any Town Fire Department. Relate who the department is and the value it brings.

- We currently have A firefighters, B emergency medical technicians, and C support personnel.
- Our coverage area includes X.
- Last year we responded to Y number of calls.
- In 2022 we donated $\$Z$ in service provided to this community.

Slide 3: Offer examples of problems you solve. Did you know we do all of the following?

- Provide home safety checks
- Offer a free cardiopulmonary resuscitation (CPR) course
- Sell address markers
- Install smoke alarms

- Teach entire programs on home safety to children
- Can provide a hands-on demonstration at your next fair or event
- Conduct inspections to ensure that your business is safe for your employees and customers

Slide 4: The ask. Any Town Fire Department is looking for volunteers.

- Community members to participate in our citizen fire academy
- People to attend our open house to find out what it is like to be a firefighter
- Individuals interested in serving as fire department members (Don't worry, we will provide the training!)
- Support at our next pancake breakfast or steak dinner

Slide 5: How can we help you?

- How can we help your organization?
- What can we do to form a collaborative partnership with your organization?
- Do you have questions?

Review Questions

1. Identify community engagement opportunities that your department can attend to _____ about what your fire service organization does and the impact that your volunteers have on your community.

2. What is a canned presentation?

Discussion Questions

1. What groups in your community can you reach out to with a request to make a presentation?

2. What are ways the fire department can help organizations within your community?

5

UNDERSTANDING THE CURRENT PROBLEM WITH THE U.S. VOLUNTEER FIRE SERVICE

There is a seat for every "ass." Find out the strengths and the weakness[es] of members and involve them.

—Stephen Marsar, battalion chief
Fire Department of New York (FDNY)

The Decline of Volunteer Firefighters

With all the value volunteer firefighters bring, it is hard to grasp that we could have a problem, but we do have a decline we cannot ignore: The number of volunteer firefighters has declined by 12% since 1984.[1] The downward trend in volunteering over the past several decades has created major challenges for current volunteer fire service organizations. However, this decline in volunteering need not lead to a doom-and-gloom mindset. Yes, we have a shift in volunteers, and yes, we need to examine why. This does not mean that we are destined for extinction.

Increasingly, we see in the headlines that volunteer fire departments are merging, transitioning to a combination department, shifting to a career department, or closing their doors. The National Volunteer Fire Council (NVFC) reported that the number of volunteer firefighters reached a new low in 2017.[2] Although the number of volunteers has declined, the need for volunteers has increased: Over the past 30 years, the volume of calls has tripled, with an increase in EMS calls being a contributing factor.

These challenges are not unique to the fire service. The largest decline in volunteering history occurred during the COVID-19 pandemic.[3] Volunteering through organizations dropped from 30% in 2019 to 23.2% in 2021.[4] Although volunteering through traditional methods declined, almost 51% of Americans reported stepping up to informally help someone in their community between

September 2020 and 2021.[5] Importantly, the declining trend occurred across states, genders, races, and ethnicities (fig. 5–1). This decline in attracting and retaining new recruits could have negative impacts, both financial and nonfinancial, within communities and is therefore a problem that should not be ignored.

The Financial Impact of Volunteer Firefighter Turnover

Volunteer turnover negatively impacts fire departments both financially and cohesively. With the average annual salary of paid firefighter being $57,120,[6] transforming from a volunteer to a paid fire department is not an option for some localities. This average salary does not include training, administrative costs, or fringe benefits. Most volunteer fire service organizations are not financially capable of adding salaries as a method to recruit and retain fire and EMS personnel. We need to be cognizant of our members' needs and the causes for disengagement to avoid costly turnover.

Failing to address firefighter turnover can lead to financial hardships for the organization. The required training and protective equipment costs associated with a firefighter moving from recruit to active duty are considerable. The NVFC reported in 2016 that the cost to outfit and train a firefighter is approximately $27,095.[7] With only one-third of all volunteers making the commitment to serve for another year, implementing strategies to address volunteer turnover should be a high priority for financial stability.

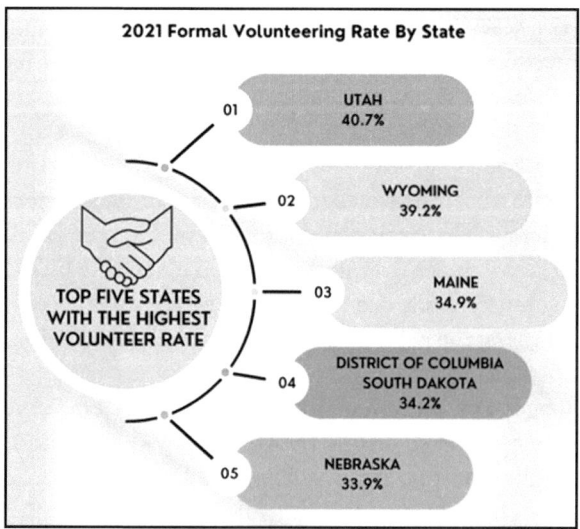

Figure 5–1. Infographic showing states with the highest rate of volunteers

The Indirect Cost Impact of Volunteer Firefighter Turnover

Fire service leaders need to understand that the cost of turnover goes beyond a dollar figure. In addition to the negative financial impact, high turnover rates can lead to weak organizational cohesion, poor morale, and low productivity. Volunteers are not required—by contract or for pay—to stay in a relationship with a fire service organization and can leave when they feel dissatisfied. Volunteer firefighter turnover can negatively impact the growth of the organization. Turnover can lead to the demoralization of the remaining volunteers and negatively affects the stakeholders the fire department serves.

Existing members can feel dissatisfaction during high turnover because they feel like they are constantly training members who inevitably leave. Nevertheless, the time they spend orientating a new member is valuable. This frustration can cause seasoned members to not make full investments in training new people because of repeated past turnover.

Have you talked with your members on how to cope with feelings tied to turnover? Are you aware of how turnover can impact future volunteer orientations? What methods have you deployed to ensure that all new volunteers are being fully trained and mentored?

Generational Impacts

While it is easy for some fire service leaders to blame the younger generation for the decline in volunteerism, data tells us that teenagers formally volunteer at almost the same rate as do members of older generations[8] (fig. 5–2). The question should be, Why are we, the fire service, unable to engage this population willing to serve their communities in other capacities?

The NVFC has reported a trend in the current demographics of fire service volunteers of reduced participation by the younger generation. This trend is alarming in comparison with the aforementioned national volunteer rates, with teenagers having a higher rate of volunteer service than even young adults 20 to 24 years old.[9]

The recruitment of younger generations in the volunteer fire service is not the only discrepancy in volunteer trends. Women volunteer in their communities three times more than men do,[10] but the NFPA reported in 2019 that only 10% of volunteer firefighters were female.[11]

Are you aware how many volunteers you have from different generations? What are you doing to bust generational myths within your organization? What trainings have you engaged in or provided to gain an understanding of generational needs?

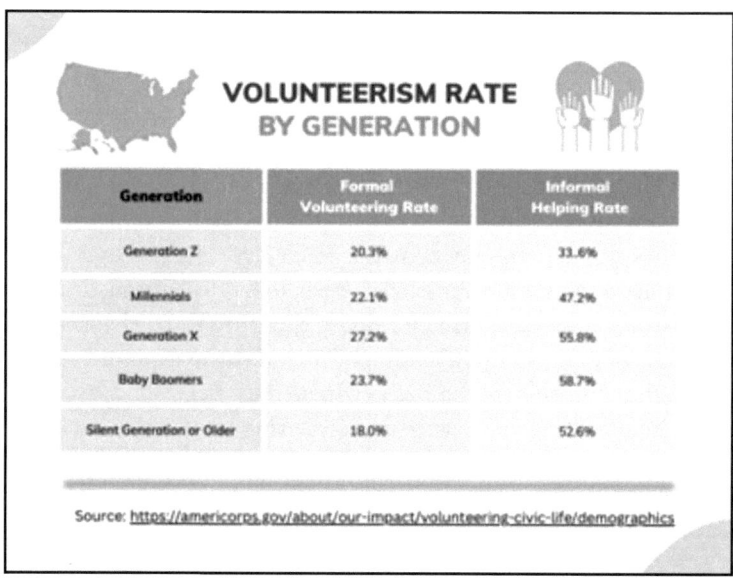

Figure 5–2. Chart showing volunteerism rate by generation

STRATEGY STOP

Communicate the generational impacts within the fire department.

Communicating generational impacts to fire department members requires a thoughtful approach to ensure understanding and engagement across different age groups. The following strategies can assist with effectively communicating the generational impacts:

- *Use diverse communication channels.* Embrace a variety of communication channels to reach all volunteers. These may include in-person meetings, phone calls, emails, text messaging, newsletters, social media, and video messages.
- *Be inclusive in language and content.* Avoid jargon or references that may be specific to a particular generation. Use inclusive language and content that resonates across generational groups.
- *Provide context to all volunteers.* Offer historical context and background information to help volunteers understand the discussed generational impacts. Explain how events from the past have shaped the fire department's present and future.

- *Visual representation.* Provide visual aids, infographics, or timelines to visually illustrate generational impacts. These can make complex information more accessible and engaging across generations.
- *Encourage two-way communication among volunteers.* Create opportunities for dialogue with and feedback from volunteers. Encourage volunteers to share their perspectives and experiences. Two-way communication can foster a sense of inclusion and promotes open exchange of ideas.
- *Emphasize common goals.* Highlight shared department goals that transcend generational differences. Emphasize how working together across generations can strengthen the impact the department has.
- *Storytelling.* Convey generational impacts through storytelling. Personal narratives can make the message more relatable to and emotionally resonant with volunteers. Be sure the stories you tell have a call to action that is clear to understand across generations.
- *Have flexibility in engagement.* Provide options for volunteers to engage with the information in ways that meet their preferences for receiving information. Some may prefer written materials, while others may prefer videos or group discussions.
- *Recognize and value contributions.* Acknowledge the contributions of volunteers from different generations. Celebrate their achievements and highlight how each generation's efforts have contributed to the fire department's growth.
- *Cultivate a culture of learning.* Promote a culture that embraces continuous learning and professional growth within the fire department. Encourage volunteers to be curious about generational impacts and engage in discussions to gain insights from one another.
- *Respect differences.* Acknowledge that generational perspectives will be diverse and these differences must be respected. Viewpoints from varying generations should be valued and welcomed.
- *Lead by example.* Demonstrate cross-generational collaboration among volunteers and leadership. Fire service leadership should embrace inclusivity and encourage others to follow suit.
- *Reinforce common values.* Emphasize the fire department's core values and use those values to unite members across generations. Reinforcing these values helps maintain unity and a sense of purpose.

Organizational Impact of Turnover

Firefighter turnover can be damaging to the organization's reputation and can negatively impact external stakeholder relationships. When firefighters leave, institutional knowledge leaves with them. Relationships with community contacts that are not properly managed can suffer. Community partners will no longer know who to call and can become dissatisfied with the frequency of change in faces. In turn, this frequent change can leave external stakeholders with the impression that the organization is dysfunctional.

High turnover can reduce organizational expertise owing to a lack of on-the-job experience. Fire service leadership must examine and address the indirect costs of employee turnover. When an employee establishes a long-term relationship with an external partner, they know the unique needs and resources the partner can offer. When volunteers leave, a knowledge loss can occur.

Another indirect cost tied to turnover is the sacrifice to employee development. The shortage in staffing often means leadership is too busy filling the roles of the missing volunteers to focus on developing remaining employees. Trying to move the department forward during high turnover is like trying to finish a race with a leaking tire: You will eventually get to the finish line, but your team will be exhausted from constantly needing to pull over to put air in the tires—and you most likely will come in dead last. Ignoring turnover and focusing only on recruitment comes at a high cost.

How are you protecting institutional knowledge? What strategies have you implemented to boost morale during turnover? Have you allowed for open and honest communication about how seasoned members are feeling about turnover, and have you shared positive coping strategies to help them deal with their feelings?

The Impact of COVID-19 on Volunteerism

In addition to the decline in volunteers over the past decade, the COVID-19 pandemic created a new problem for the volunteer fire service: The Great Resignation. This was a trend among Americans who, after the pandemic, left organizations where they didn't feel valued, that had a negative impact on their mental or physical health, and that interfered with social and family time. Our first-responder community was not an exception, and the added stressors and exhaustion led to an increased risk of turnover among volunteer firefighters. Thus, some volunteers chose to exit the industry because of the perceived risks and psychological toll.

In recent times, the trend of Great Resignation has shifted to what is known as Great Reshuffling. The U.S. Census Bureau describe this phenomenon as workers quitting and entering the labor force in an entirely new industry.[12] The willingness to explore a new career field should be a trend volunteer fire departments should capitalize on. Individuals looking for a change in vocation can obtain free fire and EMS training and experience that can translate to a paid profession.

CASE STUDY: THE GREAT RESIGNATION

Do you need to quit or just take a leave?
For transparency, I was part of the Great Resignation because of COVID-19 impacts. In March 2020, when the pandemic was shutting down the country, my husband, who works as an emergency department nurse, developed COVID and fell ill. He was the first person among his coworkers to get COVID and the person who made the pandemic real for those who knew us. His health quickly declined, and he ended up on a ventilator. While our kids and I were isolated at home, the love and support I felt from fire service members across the country was strong. We were one of the fortunate families: my husband did recover after almost a week on the ventilator and came home. However, the mental toll of almost losing him and being alone while in crisis had a strong impact on our kids and myself.

Two months after his illness, I quit the fire service. I did not want to miss a moment with my family and needed to work on my mental health. I thought that this was my retirement and that it was permanent. I even shared with others that I was retiring from the fire service. However, during my time away, one seasoned officer kept in contact with me through text messages on holidays and special events in my life. He made me feel like I was still valued even though I was not actively serving.

After almost 2 years away, the chief asked if I was ready to return, and I was reinstated within days. Retirement was not what I needed; I needed a leave of absence to address my family's needs and to work on myself. The fire service doesn't always do a good job of letting members take a long-term leave of absence and then welcoming them back. Historically, when members step away, we haze them for not showing up instead of asking them how we can support them while they take the necessary time.

This is the text message I received on October 28, 2021, 16 months after my resignation:

> Hi Candice and Happy Anniversary to you and your honey. Praying for many more. God bless.

—*Daryl Doyle, former chief and current safety officer, Winona (OH) Fire Department* (fig. 5–3)

Thoughts to Consider

- Would allowing long-term absences lead to a greater return on investments in people?
- When members leave, do we leave the door open for them to come back, or do we alienate them for abandoning ship?
- How do we show support to members when they have personal struggles? Do they know that we want to support them?

Figure 5-3. Candice McDonald with Chief Doyle

Volunteer Quiet Quitting

The phrase *quiet quitting* is another way of saying disengagement. The Washington Post reported this trend following the Great Resignation.[13] This trend is not about quitting, but rather not going above the required engagement to ensure that time is left for family and social life. It is about setting boundaries, not being a yes-man, and limiting contact hours. Even in the workforce, millennials (39%) and Generation Zers (32%) are picking jobs based on the ability to maintain a satisfactory work-life balance.[14]

Volunteers who quiet quit put in the minimum effort to stay involved. They stay silent at meetings, stop volunteering to take on duties, and run minimal calls. These individuals often lack the enthusiasm for being a volunteer firefighter. A Gallup poll in 2022 showed that 50% of workers in America qualified as quiet quitters.[15] These are workers who are psychologically detached from the job. The prevalence of quiet quitters among Generation Z and younger millennials is extremely high.

Have you noticed an increase in volunteer quiet quitting since the pandemic? Has there been a shift among a certain demographic of your volunteers? Have you talked to your membership about their stress levels or trying to find a satisfactory work-life-volunteer balance?

STRATEGY STOP

Prevent Volunteer Quiet Quitting within the Fire Department. Preventing volunteer quiet quitting, where volunteers disengage silently without expressing their concerns, is essential for maintaining a motivated and committed volunteer base. Here are some strategies to address quite quitting:

- *Regular check-ins.* Fire service leaders need to connect with their membership to prevent a decline in enthusiasm for the job. Schedule regular one-on-one check-ins with volunteers for open discussion of their experiences and to address concerns, allow for feedback, and identify areas where they need assistance to be successful. Create an open and supportive environment where volunteers feel comfortable sharing their thoughts.
- *Exit interviews.* Fire service leaders should conduct exit interviews with volunteers who decide to leave to identify the internal and external barriers volunteers are experiencing that have led to their decision to leave. This process can provide valuable insights into the reasons for their departure and identify areas where improvement is needed.

- *Volunteer surveys.* Not every volunteer will feel comfortable providing candid feedback with fire service leaders. Implement anonymous paper or electronic surveys to gather feedback from volunteers about their experiences, challenges, and suggestions for improvement. Use the survey results to identify patterns and address common concerns.
- *Clear expectations.* In studies, younger generations have reported that workplace expectations are not made clear to them. It is important for fire service leaders to set clear expectations from the beginning, including the roles, responsibilities, and time commitments involved in volunteering. Clear communication helps prevent misunderstandings and unmet expectations.
- *Recognition and appreciation.* Fire service leaders who recognize and appreciate volunteers for their efforts and contributions create an environment in which volunteers feel valued. Celebrate volunteer achievements and acknowledge the impact of their donation of time to boost motivation and commitment.
- *Skills matching.* The fire service does not do a good job matching skill sets with the desire to serve. Ask about skill set and match volunteers' skills, interests, and expertise to appropriate tasks and projects within the fire department. Volunteers who are engaged in activities that align with their strengths are more likely to stay committed. It is important to ask volunteers how they want to apply their skill set. Just because someone works as an accountant by day does not mean they want to take on the department financial records as a duty; the fire service might be their escape from numbers.
- *Training and development.* Fire service leaders need to have conversations with their members about what they need in order to grow and areas where they want to develop. Offer ongoing training and skill development opportunities to help volunteers feel valued and grow in their roles. This also enhances their sense of competence and confidence in their contributions to the fire department.
- *Team-building activities.* Organize team-building events and social gatherings to foster a sense of community and camaraderie among volunteers. A strong sense of belonging can reduce the likelihood of quiet quitting. These activities should be diverse and meet generational needs.
- *Flexibility and inclusivity.* Fostering a culture of flexibility is important. Be flexible with volunteer schedules and recognize the diverse needs of volunteers. Accommodate flexibility to allow volunteers to

meet personal commitments in order to retain volunteers who might otherwise leave owing to scheduling conflicts.
- *Continuous improvement.* Making it known that you want to foster process improvement can lead to a culture of innovation. Continuously evaluate and improve how you are managing volunteer engagement. Involve volunteers in decision-making processes to ensure that their voices are heard and valued.
- *Conflict resolution.* Fire service leaders who ignore conflict put their departments at risk for high turnover. Address conflicts or issues among volunteers promptly and constructively. Encourage open communication and provide a safe space for volunteers to express their concerns. Provide volunteers with a clear path for conflict resolution and the training on how they can resolve conflict effectively.
- *Leadership support.* Offering volunteers a supportive environment is key for retention. Ensure that department leaders are trained in volunteer management and equipped to support and engage volunteers across generations effectively.

By implementing these strategies, fire departments can create a positive and supportive volunteer environment, where volunteers feel valued, engaged, and motivated to contribute their time and efforts, reducing the likelihood of quiet quitting.

What strategies have you actively used to increase morale? How are you dealing with feelings tied to turnover? How do you foster communication and a flow of feedback?

Review Questions

1. _____ is a trend among Americans after the pandemic, in which individuals leave organizations where they don't feel valued or where they feel the organization harms their mental or physical health or interferes with social and family time.

2. Women volunteer in their communities _____ times more than men do.

3. Fire service leaders should schedule regular one-on-one check-ins with volunteers to _____.

Discussion Questions

1. What trends have you noticed in volunteer recruitment and retention since the COVID-19 pandemic?

2. What is your fire department doing to collect information from those who resign or stop showing up?

6

FOCUS ON YOUR STAKEHOLDERS

You can either make excuses and remain status quo or you can commit to the work that it takes to create change within your organization. The choice is yours!

Fundamentals of Evidence-Based Tactics

As you progress forward in this book, you will see a trend in repetitive behaviors. This trend represents the *fundamentals*. The fundamentals are strategies—evidence-based tactics that can have a significant impact on volunteer retention and recruitment. These are meant to be tweaked to fit *your* local needs. The more of these fundamentals you implement—and the more you repeat them—the more you will improve your efforts at retention and recruitment.

Fundamentals of Stakeholder Theory

It would be to my detriment as a researcher if I didn't introduce you to the one theory that underlies every strategy in this book. This proven theory is not difficult to understand—in fact, you might already be embracing it within your organization. The strategies offered here are based on the conceptual framework of Robert Edward Freeman's stakeholder theory. For those new to the term, *stakeholders* are internal and external groups or individuals who can impact organizational success, with the organization also having an impact on the stakeholder. Internal stakeholders include fire department volunteers and leadership, while external stakeholders include those external parties the organization has an impact on, such as the community.

Stakeholder theory states that leaders can maximize how an organization performs by meeting the needs of those with a stake in the future of the organization. Stakeholder theory offers a possible explanation for organizational success and stakeholder support because of its premise that leaders base their actions on the interest of stakeholders more than on self-serving agendas. Stakeholder theory has been used by researchers in disciplines such as public policy, law, and health care. It has been identified as an appropriate method for bridging strategies to stakeholder interests to create value for all involved. Researchers using stakeholder theory look to improve the effectiveness of an organization by developing and maintaining symbiotic relationships with internal stakeholders. By meeting the needs of the stakeholders, fire service leaders can maximize organizational performance.

Becoming a stakeholder-focused leader takes work and self-awareness. This requires accepting various truths about yourself and being willing to do the work to transform. We all come with leadership opposites—the leader devil and angel sitting on your shoulders as you make decisions. These comprise the balance of the good and the bad, or struggle, traits that we each possess as humans. Certain situations and personal moods can determine which trait comes forward.

It is important to take a self-inventory of your *whys* and *whos*. Why do I react this way? Why don't I react this way? Why are the decisions I am making not effecting change? Why are the decisions I am making motivating my followers? Who is the leader I need to become that my fire department and community deserve? Who can I trust to give my honest and candid feedback to?

Freeman's stakeholder theory claims that both internal and external stakeholders have an investment and benefit in the organization. Relationships between stakeholders and the organization can be beneficial for developing strategic resources for business success. Integrating stakeholders early in organizational strategic planning and decision-making can assist with the elimination of barriers to social initiatives. Under stakeholder theory, a fire service organization should serve all individuals who have a stake in the organization, and the continued existence of the organization is dependent on the support of the stakeholder group.

Stakeholder theory can assist fire service leaders with exploring strategies for retention and recruitment. Understanding this theory grants us a different perspective of the world, to solve problems blocking scientific progress and identify and implement new solutions. Volunteer turnover is not new; it has been a focus of interest to fire service for over a decade. Stakeholder theory is one way for volunteer fire service organizations to improve their organizational effectiveness to recruit and retain volunteers. Applying stakeholder theory improves the functioning of an organization by developing strong relationships with internal stakeholders. Stakeholder theory focuses on creating value for stakeholders to manage an organization.

CASE STUDY: FINDING TEACHABLE MOMENTS

I am *not* the smartest person in the crowd.
In the 8th grade, I had a guidance counselor tell my mom and me that college was not an option for me and that I would be lucky if I graduated from high school. As a first-generation college graduate, I spent years trying to prove to myself she was wrong. I would immerse myself in a topic until I felt I was an expert. I would sit for every certification I was eligible to test for. This led to being certified as a chemical dependency counselor, social worker, insurance agent, adoption assessor, firefighter, emergency medical technician, continuing education instructor, fire inspector, project manager, law enforcement officer, infrastructure protection specialist, physical security specialist, lean six sigma, Traffic Incident Management instructor, fire and life safety educator—and I even have certification from the Defense Nuclear Weapons School.

In my mind, I was becoming educated and growing my confidence. What I failed to realize is something Dr. Denis Onieal, former deputy U.S. fire administrator, later taught me. In a candid conversation we had, he instructed me to not be the smartest person in the room. He reminded me that conversations should be about "we" and not "me." He taught me to be quiet and listen, to give others the space to speak up and generate new ideas. He was not afraid to share with me that I was coming across in a way that didn't foster growth in others and instead seemed self-serving.

This was a teachable moment. This is also one lesson that I must work on often. It is not because I want to be viewed as the smartest person in the room; it is because I fear being viewed as the ignorant student my 8th-grade guidance counselor labeled me as. Having all the knowledge does not make you a good leader or motivates others. A good leader understands the value of providing stakeholders with a platform to develop and see themselves as experts. When we sit back and listen, we can learn so much from others.

Thoughts to Consider

- How does your past influence you and your leadership style?
- How can you prevent your personal issues from transferring over into the workplace?
- What are ways you can create a safe space for others to share their knowledge?
- How can you step behind others to let them be recognized?
- Who in your life can you ask for candid feedback on how you can improve stakeholder relationships?

Limitations of Stakeholder Theory

Importantly, with all theories come limitations. One limitation of stakeholder theory is that it doesn't account for the number of resources a fire department may need to fulfill all their stakeholders' expectations. These resources can be tangible and intangible.

Many fire departments lack the financial resources for a recruitment and retention budget. The fire service depends on external stakeholder support to obtain financial support for resources. This requires fire service leaders to foster relationships outside of their departments. Fully engaging external stakeholders in the daily operations of the fire service is one method for leveraging donations and taxpayer support for needed resources. If external stakeholders understand your whys, they are more likely to offer help and develop solutions. Some external stakeholders may have something intangible to contribute to the fire department that can meet the needs of internal stakeholders. It is important for fire service leaders to find creative ways to meet the needs of their stakeholders.

STRATEGY STOP

Secure memberships and passes for your stakeholders to meet their needs.

Leveraging external stakeholders to secure gym memberships, discounts, and free or discounted admission to certain places is one method to keep volunteers satisfied. The Delaware Volunteer Firefighters Association worked with elected officials to reward internal stakeholders who responded to at least 20% of the department's annual alarms or crew calls as a nonpaid firefighter, volunteer emergency medical technician, or life member of a Delaware volunteer fire department with a surf fishing vehicle permit. This permit allows the holder to drive on the beach, and only so many are issued the public each year. The public pays $90 for the permit, whereas the volunteer firefighter who meets these criteria receives it for free. By providing this perk, the Delaware Volunteer Firefighters Association served the need of their volunteers, not their own interests.

Fundamentals of Trust

A lack of transparency results in distrust and a deep sense of insecurity.

—attributed to the XIVth Dalai Lama

Most in the working population can probably recall a leader they could not trust. Trust is fundamental to every relationship. It is important to ask yourself the following questions:

- Can I be trusted, and am I viewed as a trustworthy leader? The process of trust starts with learning to trust yourself.
- Are my leadership motives based on the needs of my stakeholders or my own?
- Do I have a history of letting myself down?
- Do I make commitments to others that I can't honor?
- Do I fail to follow through?
- Do I overpromise and underdeliver?

Your followers will mimic your words and your actions. Being a trustworthy leader requires continual work and daily maintenance. Building trust among stakeholders can be a complex process. The word *trust* is invoked even when it is not always practiced. For real trust to occur among stakeholders, there must be *leadership trust* and *trustworthiness*.

Leading with trust requires an individual to be trustworthy. Being a trustworthy leader means being someone your followers view as dependable, reliable, and deserving of trust or confidence. Trust involves a relationship between two people and is not a solo act. Trust leads to positive collaboration and teamwork. Trust can contribute to improved efficiency, influence ethical decision-making, and increase loyalty, which will reduce turnover.

A trusted leader is viewed as a person who is dependable. Trusted leaders focus on the needs of the stakeholders instead of being self-absorbed. Trust is created through interactions with others, not motivational words. Trust starts with listening; being a strong listener can drive trust and influence.

Trust is a two-way process. Trust *reciprocity* is key: Trustworthy leaders who trust their volunteers are more likely to produce trustworthy organizations. When followers know that a leader puts the interests of their stakeholders first, the trust meter starts to rise (fig. 6–1).

QUALITIES OF A TRUSTWORTHY LEADER: HOW DO YOU MEASURE UP?

- I set attainable goals and empowers those I lead to grow and develop their leadership skills.
- I allow individuals to do meaningful work instead of mundane tasks.
- I openly own my mistakes, apologize when I am wrong, and share my plan for not repeating the mistake.
- I am transparent, even when the truth is uncomfortable share.
- I am candid and speak up for the betterment of the organization and the individuals I lead.
- I refrain from and redirect gossip.
- I focus on empowering weak members of the team.
- I create a culture of communication and listening, even when the topics might be uncomfortable to hear.
- I am open to feedback from those I lead and openly share that I value self-improvement.

Figure 6–1. Qualities of a trustworthy leader

As you move toward becoming a trustworthy leader, think about the questions Ben Franklin challenged us with:

- What good shall I do this day?
- What good have I done today?

> **STRATEGY STOP**
>
> **Use the trust creation process ELFEC to build trust as a leader.**
> Creating a culture of trust is a new strategy for many fire service leaders. Business leaders Charles Green and Andrea Howe have written that trust doesn't just happen; rather, it is created through conversations at the individual level.[1] They also developed a *trust creation process* using a five-step model. This process for creating trust among stakeholders is referred to by its initials, *ELFEC*:
>
> - *Engage.* "Let's talk about…"
> - *Listen.* "Tell me more…"
> - *Frame.* "So the issue is…"
> - *Envision.* "Let's imagine…"
> - *Commit.* "I suggest…"
>
> **Putting ELFEC into Action**
> A volunteer is unable to meet their required training hours to remain active. The fire chief engages ELFEC:
>
> - *Engage.* "I understand there has been some difficulty with meeting the required duty hours. Can you share what some of the barriers are?"
> - *Listen.* "I hear what you are saying. Can you tell me what you think is causing that?"
> - *Frame.* "It sounds like what you are saying is you are being forced to pick between being a good dad and attending your son's soccer games or coming to training."
> - *Envision.* "If we can make an alternative arrangement for you to meet with Capt. Jane on Saturday morning to make up the training, would that help you meet your training commitment?"
> - *Commit.* "What I can do, since Saturday doesn't work, is meet you on your free Friday night to go over the training with you. Can we do that?"

Fundamentals of Influence

The ability to influence is a key skill for fire service leaders. We know that even the wisest person can offer advice that others will not accept. A leader must earn trust and the ability to influence to effect change. Influencing others is not about convincing them you are right; instead, it is about appealing to their intellect and motivating a desire to change. Leaders who are both reliable and viewed with a deep expertise in their industry have a higher level of influence.

Transparency is another key factor for influencing others. Transparency within an organization strengthens collaboration and fosters trust. As Dorie Clark recently wrote in the Harvard Business Review, "Transparency is brand insurance. Transparency is the new leadership imperative."[1] Organizations that foster transparency are also known to have higher levels of innovation for process improvement and problem-solving.

Effective leaders understand they cannot command, but rather must encourage and motivate change. *Influence* is having the ability to have an effect on the behavior of others. Some leaders are born with a charismatic ability to inspire and influence stakeholders, and others must practice influencing skills.

STRATEGY STOP

Use the following strategies to build influence within your organization:

- *Establish relationships with stakeholders.* Take the time to get to know the names of those you volunteer with and their family members. Send handwritten cards on personal occasions they celebrate (birthdays, anniversaries, graduations, etc.). Ask questions and listen to get to know your stakeholders. You might be wondering how you are supposed to remember all of these details about every volunteer. Hair stylists have mastered this skill and have developed a technique that is easy to implement. They create an index card—their file—for each client and write pertinent information on the card to recall for the client's next visit (fig. 6–2). Each time they have a conversation with a client, they document key moments and people on the card. They review these cards to recall the personal details. People are influenced easier by those they have a relationship with. Fire chiefs can make notes of key things about their volunteers and set a calendar reminder to ask follow-up questions. For example, is one of your volunteers up for a promotion at work, getting married, having a grandchild, or graduating from college?
- *Be prepared and respect time.* Arrive early to scheduled events, meetings, and trainings. Volunteers should not have to wait for you to show up. Prepare training plans and meeting agendas in advance. Start on the set time and end on time. Even if only two people show up on time, respect them enough to start the training or meeting on time. Eliminate distractions during scheduled meetings by silencing

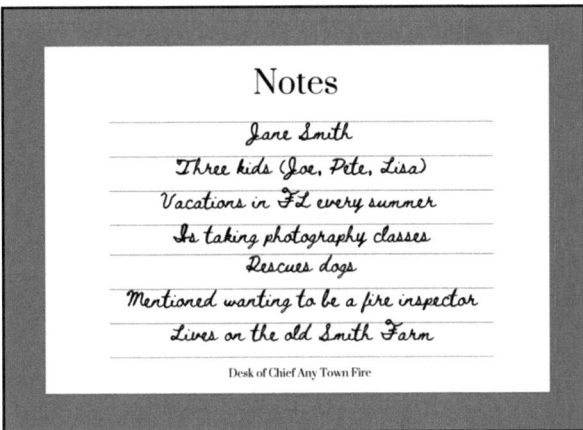

Figure 6-2. Example index card containing pertinent information to be kept on file

your phone and refraining from looking at the screen. When a volunteer feels a relationship is built on respect, they are more likely to be influenced.
- *Be transparent.* Volunteers do not want to get caught in the middle of uncommunicated change or drama. Transparency is another key factor in building trust within a team. Take the time to be transparent and communicate decisions and issues. Ensure that your volunteers understand the why behind actions and allow them an opportunity to ask questions. When transparency is constant within an organization, there is less chance that doubt will influence volunteer behavior.
- *Practice focused listening and communication.* Listening is a core communication skill—even more critical in today's digital world. Poor communication and failure to listen can lead to poor team performance. Make a strategic effort to foster communication and ensure that your volunteers know you are listening. Listening sends a message that you care and are willing to learn from others. Leaders who practice effective listening are also able to gain insight on problems within the organization and can influence change before the issue becomes a crisis.
- *Act with integrity.* While there is no law against being a terrible person, a leader who does not do what is right and proper will not be able to influence others. Know your moral compass and follow it. Speak up when you see unethical behaviors. How you react to situations has a strong impact on the influence you have.
- *Stay relevant in training and skills.* Attend and participate in trainings to keep personal skills relevant. Spend time researching industry trends and stay knowledgeable. Be ready to share industry news and

best practices during trainings and meetings. Leaders who are viewed as experts in their field have a better likelihood of influencing their stakeholders.
- *Practice resiliency.* Refrain from negativity and complaining when things do not go as planned. Actively look for solutions to overcome barriers. A good leader models the behaviors that they want to see within the organization's culture. Modeling resiliency can influence a culture of innovation and openness.

Fundamentals of Setting Goals

Many fire service leaders talk about wanting change but never set goals. It is important to ensure accountability for retention and recruitment goals. The more fire service leaders are held accountable for outcomes, the more they will invest in achieving those results. A goal provides motivation for an individual or group to reach an outcome. Goals help organizations in the several ways:

- Narrowing organizational concentration in on a focused outcome
- Shifting ideas into realities
- Breaking down larger tasks into manageable steps, or *action items*
- Prioritizing the action items needed to reach the outcome
- Providing a sense of accomplishment

Set goals that provide a focus for the organization and inform stakeholders on where they need to focus in a set time frame to allow them to prioritize tasks. One common trend that I have seen while working with fire departments across the country is a lack of structured goals to address recruitment and retention issues. Fire service leaders who set specific goals establish a clear path in the direction they want to move their organization.

Do you have recruitment and retention goals for your organization? Are your members aware of your goals? Do you have written plan with actions needed to meet your goals?

STRATEGY STOP

Conduct a backward goal-setting session.
Goals should be derived from stakeholders. One way to ensure that the stakeholders have input in the goals being set is to hold a *backward*

goal-setting session. This starts with a big-picture goal and then scales the goal into action items. Follow these steps during your backward goal-setting session:

- Ask members to develop an end goal for their recruitment and retention plan. What is it they want to achieve?
- Next, ask members to identify the milestones that need to occur to get to the end goal. How will they know when they hit a goal milestone?
- Then, ask members to identify action steps to reach milestones. What action steps will they need to reach the milestone?
- Next, ask members to identify the resources they think they will need to carry out the action steps. What resources are required to perform the actions needed to reach the goal?
- Finally, ask members to assign a responsible party to each action step who will take responsibility for leading and monitoring. Who will be in charge of each action item to make sure it is done?

STRATEGY STOP

Set and document S.M.A.R.T. goals.
This mnemonic, or memory aid, dates back to 1981.[2] S.M.A.R.T. is an acronym that spells out the following ideals:

- *Specific.* The goal will clearly define what your organization plans to accomplish. This answers the questions Why? What actions will be taken? What will be accomplished?
- *Measurable.* Criteria for measuring progress toward the goal can be identified and in turn will help you determine when the outcome has been reached. This answers the questions What data will measure the goal? How much data do you need?
- *Attainable.* The goal is within reach for the organization. This answers the questions Is the goal achievable? Does the organization have the skills and resources needed to obtain it?
- *Relevant.* The goal is realistic with a clear objective. This answers the questions How does the individual goal align with the overarching goals? Why is the outcome important?
- *Time-bound.* The goal should be timely and create a sense of urgency. This answers the question What is the established time frame for accomplishing the goal?

S.M.A.R.T. goals are strategic, focused, actionable, and easy to measure. Each goal is a statement of the desired outcome the organization wishes to achieve written in a manner that will provide clear understanding (table 6–1).

Table 6-1. Example S.M.A.R.T. goals

S	**Specific** What do I want to accomplish?	Many people ask on our social media sites for an application but never return the paper application we mail to them. Since there is an expense to mailing the application and the people requesting it are technology savvy, we want to develop an online membership application located on our website by the end of November.
M	**Measurable** How will I know when it is accomplished?	Implement an online membership application by November; and show a 20% increase in submitted applications within 6 months.
A	**Attainable** How can the goal be attained?	The website developer has indicated an online form can be developed with a specific URL (uniform resource locator) that can be shared via social media with a call to action. The project will need people to design the form, coordinate with the website developer, and serve as the point of contact to receive the submitted applications.
R	**Relevant** Does this seem worthwhile?	Increasing volunteer recruitment is a key goal for the organization.
T	**Time-Bound** When can I accomplish this goal?	For a 20% increase in submitted applications to be realized by May, the online application will need to be launched by November with a robust digital marketing campaign that includes a call to action at the designated URL.

Survey Stakeholders

It is important for fire service leaders to collect feedback from stakeholders. Surveying your stakeholders will guide creation of your strategic plan. We often spend a considerable amount of time trying to figure out what stakeholders need. Instead of asking, we make assumptions. Getting stakeholder input is crucial for making well-informed decisions, building support, ensuring project relevance, and fostering collaboration. It leads to more successful and sustainable initiatives that have a positive impact on the community or organization.

How do you collect feedback from your stakeholders? How often do you collect feedback from your stakeholders? How do you review and apply feedback collected?

Collecting feedback from fire service stakeholders is essential for several reasons:

- *Allows leadership to make informed decisions.* Stakeholders have valuable insights, perspectives, and expertise that can influence fire service activities. Collecting stakeholder input ensures that fire department decisions are well informed and consider diverse viewpoints.
- *Allows fire service leaders to identify needs and priorities.* Stakeholders can help identify the needs and priorities of the community and fire department. This ensures that the work you are performing addresses relevant issues and aligns with the collective goals of the community and fire department.
- *Allows fire service leaders to build a culture of ownership and support.* Involving stakeholders in the decision-making process fosters a sense of ownership and buy-in from volunteers. Volunteers are more likely to support and advocate for the work they are doing when they perceive that their input was considered from implementation onward.
- *Allows fire service leaders to identify problems.* Stakeholders can identify and bring forward potential challenges and obstacles that might not be apparent to fire service leaders. The early input of volunteers can be used to mitigate risks effectively.
- *Allows fire service leaders to enhance quality and relevance.* Input received from stakeholders can help fire service leaders review and direct activities to ensure that they meet the needs of the end user.
- *Allows fire service leaders to increase accountability.* Engaging stakeholders promotes transparency and accountability within the fire department. It establishes clear responsibilities among volunteers and ensures that decisions are not made in isolation.
- *Allows fire service leaders to strengthen relationships.* Involving stakeholders in the decision-making process leads to stronger relationships and partnerships. This fosters trust and collaboration, which can lead to more effective and sustainable outcomes within the fire department.

- *Allows fire service leaders to reduce resistance to change.* Stakeholder input during times of change can help address concerns and resistance. When volunteers are involved in shaping the work, they are more likely to support and champion its implementation.
- *Allows fire service leaders to meet ethical requirements.* Obtaining stakeholder input may meet an ethical requirement, especially in activities that impact vulnerable populations.
- *Allows fire service leaders to maximize resources.* Engaging stakeholders ensures that resources are used efficiently and effectively. Fire service leaders can base decisions on a comprehensive understanding of the situation, the needs of all involved, and the available resources.

STRATEGY STOP

Conduct a focus group with stakeholders to guide creation of your strategic plan.

One method to collect information on stakeholder needs is to hold a focus group. Examples of stakeholders to include in a focus group are current volunteers, past volunteers, business owners, homeowners, elected officials, and educators. Table 6–2 gives examples of questions that can be asked.

Table 6-2. Example focus group questions along with their intent

Focus Group Questions	Collection Intent
What are your top priorities, and how do you think the fire department can help?	Identify what is important to your stakeholders and what resources the fire department has that can meet those needs.
What do you feel needs to change within Any Town Fire Department?	Identify areas of weakness and untapped areas of service.
What is the biggest advancement you see within Any Town Fire Department?	Identify the most significant progress the organization has made that is valued by stakeholders.
What is the biggest frustration you see within Any Town Fire Department?	Identify the biggest frustration stakeholders have experienced for an opportunity to correct or otherwise address the issue.
What do you feel Any Town Fire Department needs to know to progress forward?	Identify information that stakeholders feel is important for fire service leaders to be aware of.

STRATEGY STOP

Conduct a survey of stakeholders to guide your strategic plan.
Developing a survey to collect feedback from stakeholders of volunteer departments requires careful planning and consideration of the information you want to gather. The following are strategies for creating an effective survey:

- *Define survey objectives.* Determine what you want to learn from the survey. Clarify the specific areas of interest, such as volunteer satisfaction, communication effectiveness, or suggestions for improvement.
- *Identify a clear target audience.* Identify the key stakeholders you want to survey. These may include current volunteers, past volunteers, instructors, community members, local businesses, or organizations that interact with the fire department.
- *Craft clear and concise questions.* Develop clear and concise questions that align with your department's objectives. Avoid leading or biased questions and ensure that each question has a specific purpose to meet your overall goal.
- *Use a mix of question types.* Implement a mix of question types, including multiple-choice, rating scales, open-ended, and demographic questions. Diversity in question types helps gather quantitative and qualitative data.
- *Test the survey.* Before administering any survey, be sure to conduct a pilot test with a small group to identify any issues with the questions, instructions, or survey flow.
- *Ensure anonymity (if needed).* If there is a history of distrust or if it may encourage stakeholders to be more candid with their responses, make the survey anonymous. This will encourage honest and beneficial feedback. Candid responses are not always easy for some leaders to hear. It is important to remember that feedback is information that can be used for continuous learning of how stakeholders view a situation. It can help some leaders connect the dots between multiple, sometimes conflicting, perspectives.
- *Set realistic completion time.* Respect stakeholders' time by keeping the survey concise. Estimate the time required to complete it and make that known up front. A short survey is more likely to receive higher response rates.
- *Provide background information.* Offer brief background information about the purpose of the survey, how the data will be used, and the benefits of participating.

- *Include an introductory message.* Start every survey with a friendly and inviting message, expressing the fire department's appreciation for stakeholders' involvement through participating in the survey.
- *Share through appropriate channels.* Choose the most relevant and accessible channels to distribute the survey across generations. This may include paper, email, social media, or website.
- *Set a deadline.* Clearly state a deadline to create a sense of urgency and encourage timely responses from stakeholders.
- *Follow-up and reminder.* Send a follow-up email or reminder to stakeholders who have not completed the survey. A gentle nudge can boost response rates.
- *Data analysis and reporting.* After the survey closes, analyze the data to identify patterns and trends. Prepare a comprehensive report that highlights key findings and actionable insights. Share the report with those who participated in the survey first.
- *Feedback and action plan.* Share the survey results with stakeholders and communicate the action plan based on their feedback. Demonstrate how their input will be used to improve the fire department.

Fundamentals of Empathy

Empathy is a skill that fire service leaders need to master. Empathy is the ability to sense the emotions of others, taking into perspective what someone else might be feeling. Empathy is important both for leadership and for recruitment and retention.

Practicing empathy can help leaders to validate the feelings of others without necessarily agreeing. For example, a township trustee might complain about the cost of a proposed recruitment plan and not see value in this investment. You could respond, "I hear you are concerned about investing a lot of money in a photo shoot for our recruitment plan and were worried about the understanding return on the investment. That makes sense." The phrase "That makes sense" expresses validation of the individual's concern while making room for a conversation.

Empathy should be a strong foundation between fire service leaders and volunteers. It fosters a positive culture that values individual contributions, promotes retention, and attracts new volunteers. Empathy is crucial for recruitment and retention in a volunteer fire department for several reasons:

- *Providing connection and understanding.* Empathy allows fire service leaders and recruiters to connect with potential volunteers personally. Understanding the motivations, concerns, and interests of potential volunteers is key for helping match volunteers with roles within the fire department that align with their skills and passions.

- *Building a culture of trust.* Demonstrating empathy builds trust between the fire department's leadership and its volunteers. When volunteers feel understood and valued, they are more likely to have confidence in department leadership and to commit, thus reducing turnover.
- *Known support and recognition.* Empathetic fire service leaders are more likely to provide the necessary support, advocacy, recognition, and appreciation that is needed for volunteers to feel valued and motivated, which can lead to continuing engagement within the fire service.
- *Addressing known concerns.* Empathy enables fire service leaders to know and address the concerns and barriers that volunteers are challenged with. This could include personal conflicts, family needs, training concerns, cultural differences, time constraints, or wellness issues.
- *Reducing volunteer burnout.* Empathetic fire service leadership is essential to recognize the signs of burnout among volunteers and implement a plan to provide the necessary resources or interventions to support volunteers.
- *Encouraging open volunteer communication.* An empathetic environment encourages open communication between all volunteers. Open communication allows volunteers to provide thoughts, constructive criticism, process improvement ideas, and feedback without fear of judgment from leadership or other volunteers.
- *Promoting volunteer inclusivity.* Empathy helps in creating an inclusive and supportive culture within the fire department. In turn, an environment that is inclusive and supportive is attractive to volunteers from diverse backgrounds and experiences.
- *Retaining seasoned volunteers.* Empathy toward seasoned volunteers entails acknowledging their contributions and commitment. Respecting seasoned volunteers in this way may encourage them to continue their service rather than considering resignation during times of turmoil.
- *Attracting new volunteers.* An empathetic approach during recruitment can make potential volunteers feel valued and appreciated. This increases the likelihood of converting them from interested to committed.

- *Enhancing team cohesion.* Empathy helps to foster a sense of camaraderie and team cohesion among volunteers. This creates a positive and supportive working environment for volunteers and leads to higher retention.

STRATEGY STOP

Transform the conversation to show empathy.
Many people confuse asking questions with showing empathy. True empathy comes in the form of statements that tell the individual you understand what they are trying to express. Forming questions after making an empathetic statement can reinforce the conversation (table 6–3).

Table 6-3. Sample conversations showing empathetic statements and reinforcement questions

Stakeholder Statement	Empathetic Statement	Reinforcement Question
I am really disappointed that your fire department has failed to recruit any members from our community.	I am sorry to hear you feel that way. It sounds like we have room for improvement from your perspective.	Would you share with our planning team some specific strategies you feel would be helpful in recruiting the target population?
I don't see value in spending money on a recruitment photo shoot with people who aren't even members of this department.	It sounds like we haven't done a good job with exploring all our options or sharing the projected outcomes of our outreach plan.	Would you mind sharing some specific ideas on how we could create recruitment photos that match the population we are trying to attract?
Your department will never change; it will always be a good-ol'-boys' club.	I can hear that you are frustrated and passionate about seeing things progress.	What do you think we could do differently to move things forward?

Review Questions

1. _____ theory states that leaders can maximize how an organization performs by meeting the needs of those with a stake in the future of the organization.

2. For trust to occur among stakeholders, there must be _____ trust and trustworthiness.

3. The five-step process for creating trust among stakeholders is called the _____.

Discussion Questions

1. What impact do you hope to make through your fire service commitment?

2. If you had to pick one fundamental to guide your leadership, what would that be? Why?

3. How has your own value system been influenced by past or current leaders?

7

UNDERSTANDING WHY FIREFIGHTERS DISENGAGE

Engaged volunteers are eager to carry out the mission of the organization. They are innovative and passionate about the contributions they are making. They are your best recruitment tool.

Volunteer Engagement

Many fire service leaders are unaware of why their volunteers choose to engage and disengage. *Engagement* relates to the volunteer's commitment to and connectivity with the organization. Fire departments with high levels of engagement among volunteers foster a culture of loyalty, retention, and high performance. The engagement level of a volunteer will also have a direct impact on the community being served and the fire department's reputation.

There are different levels of volunteer engagement. Volunteers who are engaged and those who are actively not engaged are easy to recognize: They are productively happy on the one end of the spectrum to unproductive and miserable on the other. It is those in the middle who fire service leaders often overlook—the borderline engaged. Silence among volunteers often indicates such an issue. When there is little passion or no excitement among your volunteers, you have an engagement issue.

CASE STUDY: FIX WHAT'S BROKEN

Find what's broken first—then fix it.

Last year my brand-new car's digital dashboard, radio, and windshield wipers all stopped working. I couldn't pinpoint how or where the problem started. I was just driving down the road on a cold winter day and things just stopped working. After almost 30 days at the dealership—with a new dash, new wiring, and every mechanic in the place being puzzled—it was still not fixed.

They finally flew in a lead mechanic from the car manufacturer. The lead mechanic quickly diagnosed the error, which was that the driver—me—lost an unopened beverage can under the passenger seat of the car. That can eventually froze and then exploded directly into the amplifier, which shorted the entire electrical system out. Had the dealership found the mangled can when I first brought the car in, and thought, "Hey, this is a clue," a lot of time and resources would have been saved.

When I teach recruitment and retention, attendees often want to go directly to solutions without taking the time to understand the reasons firefighters don't engage or disengage. I equate this thinking to my car repair experience: Yes, you can take your car to the repair shop and tell them to fix it without telling them the location of the issue or what behavior the car is displaying, and yes, the mechanic can eventually find and correct the problem; however, this will entail additional costs in time and resources. This is why it is important to study the reasons behind firefighter disengagement; that knowledge will help fire service leaders to develop solutions for retention and recruitment much faster.

Thoughts to Consider

- Do you have open communication such that people feel comfortable sharing truths?
- Do you discuss why people might or do disengage?
- What are some methods to identify problems related to engagement within your fire department?
- Are you actively documenting issues that arise?
- Have you talked to other fire chiefs and mentors about how to address specific issues?
- Are you repeating yourself—implementing the same solution again and again and expecting different results?

Why Understanding Firefighter Disengagement Is Key

There are two key reasons it is crucial for fire service leaders to understand why volunteers disengage: *emotional impact* and *strategy development*. First, when a volunteer leaves, it is hard for those in their department, including leaders, to not have an emotional response—leadership is left with not only an operational hole but also a feeling of abandonment, which they tend to take personally. Second, in terms of strategy, even while hearing the words "I quit" is uncomfortable for fire service leaders, those who take the time to determine the cause of the loss can gain valuable information to prevent or minimize future departures.

Collecting Volunteer Firefighter Baseline Data

Because turnover is a costly problem in the volunteer fire service, it is important for departments to collect baseline data on their organization for eventual analysis of trends, including turnover rates. Monitoring the disengagement of volunteers can reveal contributing factors to turnover. Collecting baseline data will allow fire service leaders to compare volunteer engagement before and after implementing strategies to determine if the interventions work. Quantitative data is only one way to measure turnover behavior. Collecting qualitative feedback is just as critical in telling a story.

Baseline data consists of two parts. The first is a *historical marker* that shows where the fire department has been. The second is a *forecast marker* that predicts the path the department is headed if it remains on the current course. If you are lacking historical data, don't try to re-create past data; start where you are at currently and build the department data history as time moves forward. Having at least 3 to 5 years of data for comparison is the recommended target.

Many volunteer fire departments are unfamiliar with the idea of *forecasting*. Forecasting entails analyzing historical data and determining past trends to predict the future. Forecasting questions fire service leaders can use include the following:

- Do we feel the trends that we are seeing within the department or community will continue in the same direction?
- Will the trend accelerate, slow down, or stay the same?
- Is there a chance the trend will flatten? If so, when do will that most likely occur?
- Do we think that the current trend will change? If so, what do we see happening after the trend changes?

STRATEGY STOP

Conduct a data audit to review your current methods for collecting volunteer information.
Look at the following:

- In what areas are you collecting and evaluating data?
- What do your data sets look like? For example, are you including volunteer contact details, individual hours donated, training hours achieved, calls responded to?

- What systems are you using to collect data?
- Where is your data stored?
- Who can access your data?
- Is your data kept updated and accurate?
- What are your security measures to protect collected data?
- How is your data backed up? How often? Is it automatic or manual?
- Do you print and store hard copies of data?
- Are the device's data stored on a password-protected system? Who has access? How often are passwords changed?
- Is your data stored in the cloud? If so, who owns it?
- How is your data shared?
- Are your data sets meeting your needs?

STRATEGY STOP

Conduct a demographic comparison to see how your fire department compares with the community you serve and with national volunteer engagement data.

Demographic data is helpful not only for determining where to focus recruitment efforts but also for grant reporting and ensuring equity. Understanding trends in volunteer engagement rates can help fire departments to understand the current problem—and decide where to invest recruitment and retention dollars.

National Volunteer Engagement Data[1]

Volunteer engagement rates
Rates of volunteering (in any capacity) across the United States by age group are as follows, from most to least:

- 35–44-year-olds, 28.9%
- 45–54-year-olds, 28.0%
- 20–24-year-olds, 18.4%
- 16–19-year-olds, 26.4%

Demographic data on volunteer engagement

- There is a decline in volunteering among 20–24-year-olds.
- Married persons volunteered at a higher rate of 29.9%, compared with 19.9% of unmarried persons.

- Persons with children under 18 years of age were more likely to volunteer than persons without children—31.3% versus 22.6%, respectively.
- College graduates had a higher level of volunteering than those with only a high school education—38.8% versus 15.6%, respectively.

Examine the Why

Fire service leaders can use data to start brainstorming solutions by examining the reasons that prevent a person from engaging in service. Some demands that volunteers might face include college, marriage, children, aging parents, relocating, and starting a career.

When we start to look at barriers to volunteering by life demands, we can start to develop creative methods to engage new volunteers and support existing members. The following questions may guide conversations examining barriers and solutions:

- What demands do each demographic have that might impact their ability to volunteer?
- What can the fire department do to remove any barriers tied to those demands?
- How can the fire department communicate the solutions to these barriers to the specific demographic groups?

Another way to look at the why is through a data-driven recruiting process. *Data-driven recruiting* is a method of using facts and statistics you collect on applicants and the application process to create future recruiting plans. Understanding applicant behavior through data will tell a story of what is working and what is not. Data collected during the application process can answer the following questions:

- How many applicants did you have during a specific recruitment campaign?
- How many of these applicants were not qualified?
- What was the cost of recruitment by the campaign (i.e., total cost divided by total applicants received)?
- Did you meet your recruitment goals?
- How did your applicant pool outcome compare with those of previous campaigns?
- Did you have any reduced costs associated with this recruitment campaign?

- How effective are your recruitment methods?
- Were there any changes in member acceptance rates?

STRATEGY STOP

Track recruit data to gain insight and identify trends.
In 2016 the National Volunteer Fire Council (NVFC) launched a free recruit tracking tool under their Make Me A Firefighter volunteer recruitment campaign. This is a tool that you as a fire department can use at no cost. In fact, your tax dollars have already funded the creation and maintenance of this tool.

NVFC Recruit-Tracking Tool Facts

- The website MakeMeAFirefighter.org can serve as the department's central hub for all recruitment activities (fig. 7–1).
- Fire departments can list their volunteer opportunity at http://portal.nvfc.org.
- Volunteer applications can be submitted through MakeMeAFirefighter.org.
- Fire departments are notified of the interested applicant, and the applicant's record in the system is marked "interested."
- MakeMeAFirefighter.org allows departments to monitor each stage of the application process.
- Departments can manually add prospects not received through MakeMeAFirefighter.org.

Understanding Research-Based Barriers to Retention

If you are unsure of what barriers exist locally, you can use my research as a baseline. During my research, the following barriers emerged as the most frequently reported barriers to volunteer firefighter retention:[2]

- Sleep deprivation
- Gender-specific issues
- Mental health
- Work-life-volunteer balance
- Generational factors
- Organizational climate and culture

7 | Understanding Why Firefighters Disengage 91

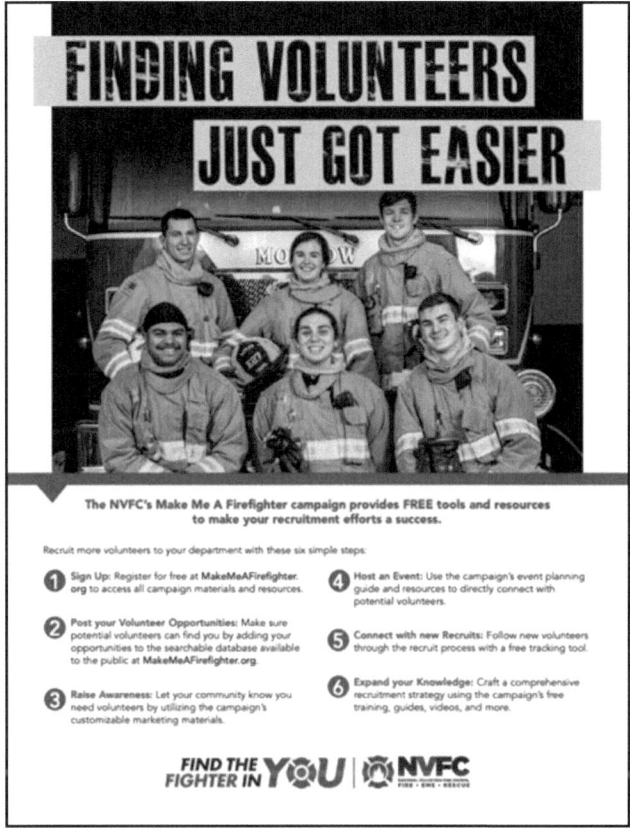

Figure 7–1. Flyer from MakeMeAFirefighter.org to encourage volunteering

These barriers (fig. 7–2) will be discussed throughout this book, along with strategies to remove them.

Conducting research—and understanding findings—is important for generating strategies and guiding decision-making. Research can help us understand the reasons behind an issue, such as why volunteer firefighters disengage. In addition to exploring the research summarized here, fire service leaders should delve into issues specific to their own communities to ensure that the strategies meet the needs of their stakeholders.

Assessing Volunteer Engagement

Gaining information for strategy development is key. When a fire service leader uses information about why valuable volunteers leave as a learning opportunity to create organizational change, corrective action to prevent future losses can be

Figure 7-2. Chart showing research-based barriers to retention

the outcome. Assessing volunteer engagement can help fire service leaders to rethink policies, practices, and operations to provide stakeholder support. Volunteer engagement information is collected and assessed to create internal changes, as a method for effectively mobilizing your volunteers, or potential volunteers to meet the organization's goals and mission.

One Canadian report has stated that the first step in assessing volunteer engagement is to ask the following:[3]

- What you hope your volunteers will accomplish
- How volunteers feel they can be engaged more effectively
- How your governing body (board, trustees, council members) can be involved within the organization

In addition, solicit stories of your volunteers in action and publish them. Then examine takeaways from these member stories.

STRATEGY STOP

Conduct an exit interview and survey to see why volunteers are leaving.

It is important to uncover the details about why a volunteer chooses to go elsewhere or stop volunteering entirely. An exit interview offers the volunteer a chance to speak openly about their experience at the fire department, and an exit survey provides data to identify disengagement trends.

Tips for Conducting an Exit Interview

- *Select an unbiased interviewer.* The person conducting the interview should be someone external to daily operations or even the fire department. This creates an unbiased atmosphere.
- *Advance preparation.* Exit interview questions should be prepared in advance, and the interviews themselves should be held in a private location, potentially outside of the fire department depending on the circumstances of why the person is leaving.
- *Offer an exit survey.* Send the exiting volunteer a written survey to complete before the exit interview. This will give the volunteer time to think about their responses. Some volunteers may be more open to the written survey than the exit interview. The exit survey can also guide the conversation during the exit interview.
- *Get it scheduled.* Be proactive in scheduling the exit interview; do not put it on the volunteer to reach out to schedule. Schedule the exit interview at a time that accommodates the exiting volunteers' schedule.
- *Listen, listen, and listen.* Actively listen and ask questions to avoid making any assumptions. Reiterate to the volunteer that you value their feedback.

Sample Exit Interview Questions

- Why are you choosing to leave the fire department?
- Would you recommend volunteering with the fire department to your friends and family? Why or why not? Who?
- What do you feel the fire department does well?
- What can the fire department do to improve how we support our volunteers?
- Do you have any specific feedback about fire department officers or others you interacted with during your time here?
- Do other volunteers feel the same way you do about the organization?
- How do you and other volunteers feel about how leadership communicates with volunteers?
- What can the department do to improve communication?
- What was your favorite thing about volunteering here?
- What ideas do you have for improvement that you wish we could have implemented while you were here?

- Are there certain members who made a positive impact on your time here? Who? How?
- What piece of advice would you give a new volunteer to be successful here?
- Would you be interested in still helping occasionally, with fundraisers, outreach, or in another manner?
- Is it okay if we stay in contact with you through email, text, or phone?

STRATEGY STOP
Conduct a volunteer engagement survey to measure volunteers' commitment, passion, motivation, and sense of purpose.
Importantly, if there is low participation in the engagement survey, that typically means there is a lack of trust between volunteers and leadership. Volunteer anonymity should be stressed and kept during the survey. Demographic information that could lead to identifying volunteers should not be included in the survey. The survey should be direct and easy to comprehend. Questions should be designed to measure the level of volunteer engagement.

Sample Volunteer Engagement Questions

Rating questions
Respond to each using the following rating scale: 1—I strongly agree; 2—I somewhat agree; 3—I am neutral; 4—I somewhat disagree; 5—I strongly disagree.

- I would recommend volunteering with this organization to a close friend or family member.
- I intend to be volunteering here in 12 months' time.
- Volunteering for this fire department makes me want to do the best job that I can.
- I feel excited about showing up for fire department trainings.
- I feel excited about showing up for emergency calls.
- I feel excited about showing up for department meetings.
- I am proud to be a volunteer with this fire department.
- I am satisfied with the nonwage benefits the fire department provides for my service.
- I enjoy volunteering with the other department members.
- I find the work we do at the fire department meaningful.
- I am familiar with the fire department's mission, vision, and values.

- Fire department leaders recognize my work and accomplishments within the fire department.
- My peers recognize my work and accomplishments within the fire department.
- Fire department officers are invested in seeing me be successful within the fire department.
- The culture within the fire department is supportive.
- The culture within the fire department fosters inclusiveness.
- I am given the opportunity to demonstrate my skills on the job.
- I am offered training to advance within the fire department.
- I see a path for career advancement within the fire department.
- I have the tools needed to maximize my potential.
- I have thought about leaving the organization within the past 6 months.

Open-ended questions

- What practices do we need to change within our fire department?
- Are there any issues with the current culture of our fire department?
- How can we increase volunteer engagement in our fire department?
- Is there anything you would like to share that would help make our fire department a better place to volunteer?

The Impact of Sleep Deprivation and Sleep Deficiency

Sleep deprivation is defined as a condition that occurs if you don't get enough sleep. In turn, sleep deprivation can contribute to *sleep deficiency*, which is a more complicated condition related to both quantity and quality of sleep, and one key contributor to sleep deprivation is getting sleep at different times of day.[4] Our bodies have internal clocks that tell us when we are ready to sleep, known as a *circadian clock*. Our circadian clocks can be disrupted by artificial bright lights, stimulants, and smartphones or, in the case of first responders, our pagers. Insufficient sleep can impact our physical health, how our thoughts are processed, and how memories are stored.

Sleep deprivation is a deterrent to the volunteer fire service. Sleep deprivation can lead to slower reaction times and increase the risk of an accident while operating a vehicle.[5] Being aware of these risks is important for fire service leaders—knowledge of them can guide future solutions to ensure a culture of safety as a priority.

Insufficient sleep caused by working at night and different shifts, such as the work performed by volunteer firefighters, can lead to *shift work sleep disorder*

(SWSD). The Cleveland Clinic reports SWSD increases the likelihood of all of the following:[6]

- On-the-job accidents and errors
- Moodiness and irritability
- Impaired social functioning
- Poor coping skills
- Gastrointestinal, cardiovascular, and metabolic problems
- Substance dependency

First responders when awakening for a call-out tend to feel groggy. This is a result of *sleep inertia*, defined by the Centers for Disease Control and Prevention (CDC) as "temporary disorientation and decline in performance and/or mood after awakening from sleep. People can show slower reaction time, poorer short-term memory, and slower thinking, reasoning, remembering, and learning speed."[7] Sleep inertia typically lasts 30 to 60 minutes but has been documented to last up to 2 hours.[8] The effects of sleep inertia are a concern for responding volunteer firefighters because they do not have the flexibility to fully wake up immediately when being called out.

Sleep deprivation in volunteer firefighters is a direct result of inadequately balancing family life, career, and the necessary requirements for being a volunteer firefighter. *Short sleep duration* is defined as getting less than 7 hours of sleep within a typical 24-hour period.[9] Working long hours may lead to sleep disturbance and psychological distress. A sleep disturbance is viewed as an actual disorder of sleep-wake schedule.[10]

Volunteer firefighters who are awoken during the night to deploy to an emergency and then report to their paying job also experience the stressor of *sleep restriction*. Volunteer firefighters report an average of 3 to 6 hours of sleep on nights of deployment.[11] Inadequate sleep also leads to poor performance and impacts safety. In addition to the nighttime call-outs disrupting sleep, work stresses of firefighting can lead to difficulty sleeping. These stressors include exposure to trauma, diseases, job-related injuries, and providing services to difficult people.

The impact of sleep deprivation extends beyond how a volunteer functions as a firefighter. Lack of sleep can negatively affect their performance at their paying job, interactions with loved ones, and overall mental health. In the long term, sleep deprivation can also negatively affect the physical health of first responders. The CDC reports that sleep deprivation is linked to serious health conditions, such as type 2 diabetes, heart disease, depression, and obesity.[12]

Fire service leaders need to develop solutions that focus on reducing sleep deficiency tied to volunteer firefighting. The National Heart, Lung, and Blood Institute considers sleep to be a basic human need—no less important than drinking, eating, and breathing.[13] The lack of sleep also puts others at risk when an individual is involved in a patient care or transportation role.[14]

STRATEGY STOP

Provide sleep health education to volunteers annually.
Although sleep is an essential function that allows both our body and mind to recharge, not everyone is educated on the importance of getting enough sleep or the proper amount of sleep the body requires. Many of us adopted the belief, "I can sleep when I am dead." It is important for fire service leaders to create programs to educate volunteers on the dangers of not getting enough sleep and provide tactics to promote good sleep to support volunteer wellness and productivity.

Educating Volunteers on Sleep

- Offer training on sleep health—for both volunteers and their families—that is hosted by a health care professional.
- Encourage volunteers to keep a sleep journal (fig. 7–3) to track hours of sleep and identify personal barriers to getting 7-8 hours of sleep.[15]
- House creditable resources about sleep in the station and online for easy access (fig. 7–4).[16]
- Ask volunteers for feedback on methods to encourage good sleep habits.

STRATEGY STOP

Implement a fire department sleep initiative to promote healthy sleep intervention(s) and collect data on the sleep culture of the fire department.

Tips from the CDC Workplace Health Resource Center[17]

- *Assessment.* What information about my volunteers and my organization can I use to select an appropriate sleep intervention?
- *Planning and management.* How will I empower and encourage my volunteers to participate?
- *Implementation.* What supports do I need to put in place for stakeholders to make the intervention a success?
- *Evaluation.* How will I measure whether the intervention is successful and sustainable?

Sleep Diary

Use this sleep diary to record the quality and quantity of your sleep; your use of medicines, alcohol, and caffeinated drinks; and how sleepy you feel during the day. Bring the diary with you to review the information with your doctor.

Fill out before going to bed	Today's date:	June 13*							
	Number of caffeinated drinks (coffee, tea, cola) and time when I had them today:	1 drink, 8 p.m.							
	Number of alcoholic drinks (beer, wine, liquor) and time when I had them today:	2 drinks, 9 p.m.							
	Naptimes and lengths today:	3:30 p.m., 45 minutes							
	Exercise times and lengths today:	None							
	How sleepy did I feel during the day today? 1—So sleepy I had to struggle to stay awake during much of the day 2—Somewhat tired 3—Fairly alert 4—Alert	1							
Fill out in the morning	Today's date:	June 14*							
	• Time I went to bed last night: • Time I got out of bed this morning: • Hours spent in bed last night:	11 p.m. 7 a.m. 8							
	Number of awakenings and total time awake last night:	5 times, 2 hours							
	How long I took to fall asleep last night:	30 minutes							
	Medicines taken last night:	None							
	How alert did I feel when I got up this morning? 1—Alert 2—Alert but a little tired 3—Sleepy	2							

NIH National Heart, Lung, and Blood Institute

* This column shows example diary entries—use as a model for your own diary notes.

Figure 7–3. Example sleep diary

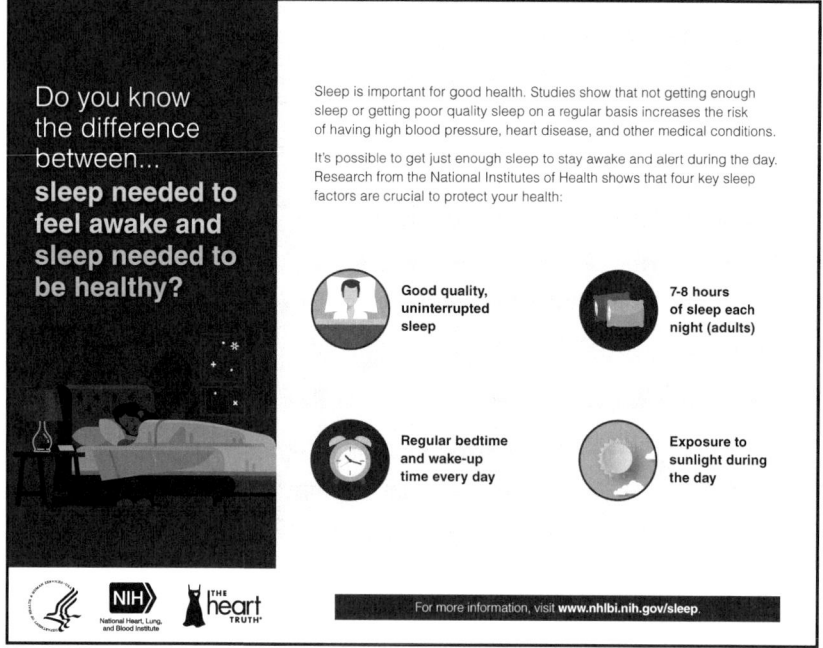

Figure 7–4. Infographic explaining the importance of sleep

STRATEGY STOP

Move to duty or credit hours versus call response percentage requirements to allow for scheduled sleep.
Moving from a percentage requirement to a duty crew requirement will accommodate the sleep schedules of volunteers. We cannot predict when emergencies will occur, which places volunteers at a disadvantage for proper sleep health when a percentage requirement model is used.

Moving to a quarterly credit hour system for membership requirements allows volunteers the opportunity to remain in good standing by making up credit hours in a quarter for a month that might have been missed owing to outside commitments. This system is also a critical strategy for work-life-volunteer balance.

Sample Credit Hour System
(Adapted from the East Franklin (NJ) Fire Department)

- *One credit.* Up to 1 hour of any firehouse activity or response to a fire or emergency medical services (EMS) call.
- *Two credits.* Up to 2 hours of firehouse activities.
- *Three credits.* Activities over 2 hours and up to 3 hours.
- *Four credits.* Duty crew.
- *More than four credits.* Will be issued by the duty officer, chief, or other designee.

Membership requirements for active regular members
Each member must accrue 35 combined fire or EMS and activity credits per quarter. Each member must also attend two general meetings per quarter. A member with a valid reason (all reasons will be reviewed by the executive board for final approval) may be excused from attending the two general body meetings per quarter. A member who is excused with a valid reason will be required to accrue 40 combined fire and activity credits per quarter. All members, regardless of status, must attend all training required to maintain fire or EMS certification.

Any member delinquent for one quarter will be put on probation for a period of 6 months and will be required to accrue 15 fire or EMS activity credits per month of the probationary period. The member must attend one general meeting per month. If for any reason the member cannot meet this requirement, approval for waiver by the executive board must be obtained. The probationary period begins immediately following the member's delinquency and will run concurrently.

All members who comply with annual requirements will start the following year with an unblemished record. If a member does not meet the above stated requirements, the member will be dropped from the rolls.

STRATEGY STOP

Implementing work-rest cycles should be mandatory.
Unfortunately, few studies in the United States have explored interventions specific to firefighter sleep, while 37.2% of firefighters report having a potential sleep disorder.[18] To reduce the negative impacts of sleep fatigue, fire service leaders can require work-rest cycles before firefighters operate equipment.

Case Study: Disaster Due to Sleep Deprivation
History is full of incidents involving disastrous accidents involving human errors blamed on sleepiness. Examples include the following incidents:

- 1979, Three Mile Island nuclear power plant accident in Pennsylvania
- 1986, Space Shuttle *Challenger* explosion off the coast of Florida
- 1986, Chernobyl Nuclear Power Plant accident in the former Soviet Union
- 1989, *Exxon Valdez* oil spill off the coast of Alaska

When I became an employee of the National Aeronautics and Space Administration (NASA), I was constantly reminded of past lessons in NASA history demonstrating the detrimental impacts of prolonged and unusual shifts. NASA places a high priority on preventing employee psychological and physiological stress and undesirable outcomes. To minimize human error factors, especially those tied to fatigue, employees were required to abide by safe work-rest cycles and shift scheduling. NASA stressed a need for considering individual circadian rhythms to ensure adequate work and sleep-rest cycles and to allow adequate time for adaptation and recovery from old to new shifts or change in time zone.

To minimize worker stress and fatigue related to time factors, NASA mandated that under no circumstances would an employee be required to work a schedule without at least 8 hours off duty between shifts—with a minimum of 10 hours off duty preferred and 12 hours or more considered optimal to accommodate employee commute time and domestic and sleep needs.

When the 8-hour period is shifted within the 24-hour day-night cycle (shift work), compensatory time must be allowed for circadian rhythms to adapt. It was noted at NASA that forward-rotating shifts—from day to evening to night, rather than counter—are easier for human adaption. NASA also required leadership to define the "standard" work period for all operations and tasks, including method of shift rotation if required, as well as breaks and required rest cycles.

What does NASA have to do with the volunteer fire service? Fire service leaders can use this example to educate volunteers on the benefits of work-rest cycles and require work-rest cycles before firefighters are allowed to engage in fire suppression or operate heavy equipment. Leaders should also have discussions with members about the root causes of past disasters and creating a culture of safety.

Thoughts to Consider

- What impacts can sleep deprivation have on high-risk occupations?
- How can sleep deprivation lead to poor judgment and human error?
- How can fire service leaders educate firefighters on the dangers of sleep deprivation?
- What can fire service leaders do to promote good sleep hygiene?

Gender-Specific Issues Related to Disengagement

Gender-specific issues impacting firefighter retention have result because the fire service has failed to fully adapt to diversity and the specific needs of the female firefighter. The fire service has been classified as a male-dominated field, with 11% of volunteer firefighters being female[19.] This statistic conflicts with the fact that women volunteer their time with non–fire service organizations three times more than men, thus showing a pool of potential candidates willing to serve their community. Interestingly, there are more female volunteer EMS professionals than paid, 38% versus 22%, respectively.[20]

Transgender volunteers have not always felt welcome or safe within the fire service. In general, more than one in four transgender people have lost a job owing to bias, and more than three-fourths have experienced discrimination in the workplace.[21] In the United States, 1.6 million individuals identify as transgender[22]—and 1.6 million is a big pool of potential volunteers!

A trans-inclusive workplace is unfortunately not yet the norm within the volunteer fire service. Many fire service leaders are unsure how to address this topic, so they avoid it. Many fire service organizations remain ill-equipped to create the policies and workplace cultures that would support trans volunteers. Here are some facts for fire service leaders to consider regarding the trans population:

- Showing respect and encouragement in the workplace to a trans volunteer as a human being does not mean that you have to change your religion or even your personal views. It simply means you are respectful to a fellow firefighter.
- Volunteers who work hard and contribute to an organization's success should never have to feel stigmatized or fearful of coming to work.
- Any form of discriminatory behavior is harmful to an organization.
- Allowed hostility and discrimination within an organization increases absenteeism, undermines commitment and motivation, and decreases productivity.
- A lack of trans-specific policies within the fire service can lead to higher turnover and even litigation.

Ill-Fitting Gear and the Impact on Female Success

One barrier impeding women from succeeding is the lack of properly fitting equipment. Many fire departments still lack proper-sized gear to outfit female firefighters. This forces female recruits to wear gear that impairs their ability to perform the tasks required to become certified firefighters.

CASE STUDY: ILL-FITTING GEAR MAY INCREASE FAILURE RATES

As a female firefighter who went through fire school with ill-fitting fire gloves, I liken wearing those gloves to wearing oven mitts while trying to carry out fireground operations. The simple task of inserting the wooden wedge to stop water flow at the top of the tower became impossible for me. The oversized, wet gloves lacked dexterity and led me to pull off the glove to complete the task. As you can imagine, removal of personal protective equipment (PPE) did not go over well with my instructor!

I had been issued a size medium fire glove to wear in fire school. Now, thanks to my good friend Tom McCoy of Fire Craft Gloves, I wear extra small fire gloves. The next time you are in a kitchen, slide on a pair of oven mitts and try to tie your shoe. That simple task will demonstrate the barrier ill-fitting gear creates.

Thoughts to Consider

- How can ill-fitting gear impact operations?
- How can ill-fitting gear impact safety?
- What message does issuing ill-fitting gear to a new recruit send?
- Does your supplier understand sizing individuals properly?

Ill-fitting gear being issued to female firefighters—or any firefighter, for that matter—is a safety liability. Note that this issue goes beyond the fire department itself. There is a need for uniform and PPE manufacturers and vendors to offer—and promote—sizes made for a woman's physique. Not all vendors offer female cut or sizes for petites (men or women). Issuing gear that leads to safety concerns because of being ill-fitted sends a message to a recruit that their safety is not valued and does not contribute to fireground operation success. Volunteers are more likely to disengage when they do not feel safe or valued.

STRATEGY STOP

Develop a gear-sharing program with mutual aid fire departments or countywide to have a wider selection of gear for new recruits to use during initial training.
This will lead to a bigger selection gear and can help to ensure that all PPE issued fits properly.

Developing a gear-sharing program among mutual aid fire departments involves careful planning, collaboration, and coordination to ensure its success. A gear-sharing program among mutual aid fire departments can enhance operational capabilities, improve response times, and promote effective emergency management across the participating communities. Follow these steps to create an effective gear-sharing program:

- *Assess department needs and resources.* Conduct a thorough assessment of the gear requirements and existing resources of each mutual aid fire department involved. Identify any gaps or surpluses in gear inventory.
- *Identify key contacts for each department.* Designate key contacts or liaisons from each fire department who will be responsible for communication and coordination throughout the program.
- *Establish a memorandum of understanding between departments.* Create a formal agreement, known as a *memorandum of understanding*, that

outlines the guidelines and conditions of the gear-sharing program. Specify details about the types of gear that can be included in the sharing program, the duration of time over which the gear can be borrowed, how gear maintenance will be addressed, how the gear will be used, how gear will be tracked, and any liability considerations.
- *Define procedures.* Develop clear procedures for requesting and borrowing gear. Specify how requests are to be made, the time frame for response to the requests, and any special considerations for fast-tracking gear lending for critical incidents.
- *Train and familiarize.* Conduct training sessions to familiarize firefighters with shared gear responsibilities to ensure that all users know how to use and maintain it properly.
- *Track and monitor.* Implement a system for tracking gear usage, including check-in and checkout procedures. Conduct quarterly inventory checks. Regularly monitor gear condition and perform necessary maintenance to ensure safety and compliance during those quarterly checks.
- *Communicate and collaborate.* Maintain regular communication among participating fire departments. Share updates on how the gear-sharing program is working, best practices, and any lessons learned for process improvement.
- *Secure funding (if needed).* If the gear-sharing program requires additional funding, explore regional grants. Regional applications that cover multiple departments make for a strong need proposal.
- *Evaluate and improve.* Periodically evaluate the gear-sharing program's benefit. Solicit feedback from users and make necessary improvements to address any challenges that arise.
- *Promote the spirit of mutual aid.* Emphasize the importance of mutual aid collaboration and reinforce the spirit of supporting other departments in accomplishing the same mission.
- *Review legal and insurance considerations.* Ensure that the gear-sharing program complies with legal and insurance requirements. Consider liability coverage and risk management strategies.

Fire Department Culture Toward Women

Another barrier for women is the culture within the fire service. Research shows that inclusion, acceptance, and friendship are all motivating factors for volunteerism. Negative attitudes and discrimination toward female firefighters from male counterparts have become a recognized barrier to female firefighter retention. Such attitudes can negatively affect the mental health of female firefighters. Female firefighters are at higher risk for anxiety, coworker conflict, and discrimination than males, all factors that may lead to disengagement.

Importantly, males are not the only ones creating an environment of incivility for new females entering the industry: Women firefighters are just as guilty. Women who were the first or only females in their departments have reported changing aspects of their personalities to be accepted within the firehouse and that they didn't want a new female to ruin what they had worked hard to establish.

When female firefighters, or any volunteers, feel inadequately accepted, they are more likely to disengage. Ongoing threats to acceptance and belonging can lead to negative effects within the organization.[23]

Dr. Sara Jahnke reported at the 2023 Congressional Fire Service Institute that of 1,773 women firefighters surveyed, 41% experienced shunning or isolation, 38% were victims of verbal harassment, and 5% reported being assaulted in the workplace (fig. 7–5). Ignoring unhealthy cultures that fail to address discrimination may lead female firefighters to experience emotional changes or emotional numbness.

STRATEGY STOP

Become educated on best practices to prevent and address discrimination, harassment, and retaliation in the fire service.
With funding from the U.S. Fire Administration, Women in Fire and the National Volunteer Fire Council (NVFC) developed the Discrimination & Harassment Toolkit, as a guide for individuals and organizations on discrimination, harassment, and retaliation (this guide can be downloaded for free from https://www.nvfc.org/wp-content/uploads/2023/04/Discrimination-and-Harassment-Toolkit.pdf). Fire service leaders should familiarize themselves with the best practices outlined in this guide, make it available to members, and hold annual trainings to ensure that all members understand the legal obligations. Trainings should include opportunities for members to ask questions and share.

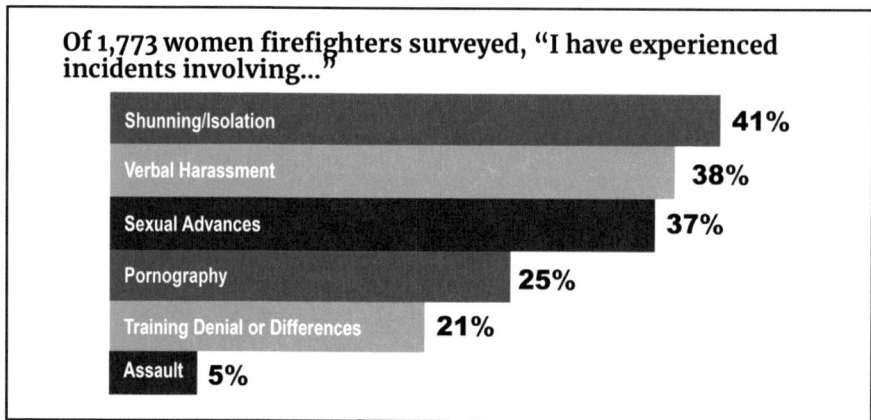

Figure 7–5. Graph showing survey responses of women firefighters pertaining to workplace incidents

Injury Rate among Female Firefighters

As a fire service leader, are you aware of the data on injury rates? The injury rate among female firefighters is 33% higher than that among their male counterparts.[24] Many female firefighters contribute to this by pushing themselves to the point of injury to avoid being viewed as weak by their male counterparts. For example, a male firefighter may ask for assistance from a peer with a task when fatigue sets in, whereas a female will not because she doesn't want to be seen as incompetent. One female firefighter shared with me that she received second-degree burns in a fire because she did not want to be the first one to retreat; instead, she waited until a male counterpart made the decision to leave to avoid having her peers think that she could not perform. Her failure to feel comfortable speaking up led to a preventable injury.

CASE STUDY:
PRIDE MAY PREVENT AN ASK FOR HELP

I Will Do It Myself

Safe lifting and moving of patients is critical to our industry. However, even in 2021, the National Emergency Medical Services Education Standards still listed it as a key area for all levels of EMS under Workforce Safety and Wellness.[25]

Because I am smaller in stature, lifting patients was an area that once gave me anxiety. I was terrified of dropping a patient. That all changed on one call when Medic and Firefighter Brandon Beeson told me that lifting was all about physics, not strength. Those encouraging words have stuck with me all these years and empowered me to approach situations with science instead of fear.

As the first female hired in the fire department I served at the time, I was determined to prove that I could do the job without the performance bar being lowered. I had been in EMS for almost 10 years and prided myself on being efficient. We were dispatched at 1 a.m. to a routine medical call; I went directly there from home, and the ambulance was already on scene. As I headed to the house, hearing over the radio that this would be a transport, I had decided to grab the power cot, which the department had invested in shortly after I was hired. This would be my first time navigating the new cot.

The entrance to the house was uphill with several rickety, old, icy steps. The power cot and I weighed the exact same amount: 125 lbs. I fell on my butt not once but twice working to get it up onto the porch. Just as I accomplished my goal, with bruising sure to come, the captain walked out and saw it. He looked at me puzzled and said, "That thing is a beast. I just get the stair chair or the reeves. I never bring it in the house, unless there are multiple people to help." I was so worried about being viewed as weak for not being able to manage the new cot that I inadvertently put myself at risk for injury.

Thoughts to Consider

- How can gender stereotypes prevent women from asking for help?
- How can department leaders and trainers educate others on the different ways tasks can be performed to achieve the same results?
- How can departments create a culture in which speaking up and asking for help is viewed as a strength?

Reproductive Risks

At the 2022 Fire Department Instructors Conference (FDIC), Dr. Sara Jahnke shared that women firefighters are 2.3 times more likely to have a miscarriage than nonfirefighters. She also reported that volunteer firefighters had a 42% higher miscarriage rate than career firefighters. Furthermore, women in protective services have the highest rate of suicide of any other occupation studied. These numbers are not only concerning but also can be a deterring factor for females exploring the opportunity to become a volunteer firefighter.

The Impact of Mental Health on the Volunteer Fire Service

First responders must be aware that the tasks that we perform are both physically and mentally demanding. I encourage departments to incorporate both physical and mental health training in the training programs for new recruits and reinforce this training with continuing education for all members on an annual basis.

—Dave Lewis, director, NVFC Maryland

Fire service leaders must become educated on mental health challenges, needs, and interventions. Unaddressed mental health issues can lead to turnover. It is key for fire service leaders be proactive in developing programs to improve retention rates. Americans, in general, are at risk for mental illness, with one in five adults living with some form of mental illness.[26] Because of the nature of the job, behavioral health risks for first responders are a concern, including post-traumatic stress disorder (PTSD), depression, stress, substance use, and suicidal ideation.[27] The psychosocial demands first responders face are tied to increased turnover and mental health issues among volunteer firefighters. Mental health issues in general can interfere with job performance, productivity, work engagement, communication with colleagues, physical capability, and daily functioning.[28]

Mental health issues among volunteer firefighters are often a direct result of the job functions they are expected to perform. The role of the firefighter has undergone a transformation—from once focusing solely on fire suppression to now responding to potentially traumatic experiences such as emergency medical calls and terrorist attacks. This shift in roles has led to an increase in vulnerability toward mental health issues in the fire service. Firefighters are repeatedly exposed to unpredictability, taking risks to personal safety, dealing with death, and being exposed to disaster. The traumatic stress firefighters face is a precursor for PTSD, substance use disorders, and other mental illnesses. One study of firefighters indicated 46.8% had suicidal ideations compared with 13.5% among the general population.[29] Dr. Sara Jahnke shared at FDIC 2022 that 15.5% of firefighters had attempted suicide.

Volunteer firefighters with untreated mental health issues are more likely to leave the fire service, and volunteer firefighters are less likely than career firefighters to seek mental health services. Although mental disorders and substance abuse are prevalent among firefighters, there is also a general reluctance among firefighters to seek treatment. Many firefighters report lack of comfort with outside mental professionals and lack confidence that outsiders could understand their experiences.

Another barrier is that mental health issues are not always apparent to health care providers. When receiving care from a doctor, patients typically report their physical problems and do not always bring up emotional health issues.[30] Even if a health care provider is aware of the mental health vulnerabilities within the first-responder community, they don't always know that a volunteer is serving in that capacity because their paid employment would be the career listed in their medical records.

CASE STUDY: SUICIDE DOESN'T DISCRIMINATE

In 2010 my cousin Debbie, a volunteer firefighter, shot herself and ended her life. This was a time when firefighter behavioral health was not being talked about. She left behind three children and several grandchildren. At the time our family didn't understand why Debbie chose this path, but over the past decade, we have become educated on the impacts being a first responder can have (fig. 7–6). According to Dr. Sara Jahnke at FDIC 2022, women in protective services had the highest suicide rate of any occupation in one study.

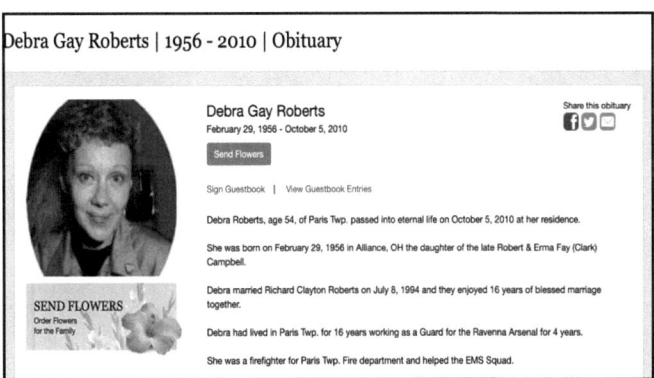

Figure 7–6. Obituary of Debra Gay Roberts

Our family's story is not uncommon. An increasing number of firefighters are dying as a result of suicide.[31] It is important for all members of the fire department to become familiar with the warning signs of suicide and know what to do when they see them. The Help Network of Northeast Ohio recommends as an easy-to-remember mnemonic IS PATH WARM?

- Ideation
- Substance abuse
- Purposelessness
- Anxiety
- Trapped
- Hopelessness
- Withdrawal
- Anger
- Recklessness
- Mood changes

If you suspect a person is at risk for suicide, seek help by contacting a local mental health professional, calling 9-8-8, or using the NVFC's First Responder Helpline (fig. 7–7). The helpline is free for all members and their families.

Thoughts to Consider

- Does your fire department have a local mental health professional available to members?
- What resources has department leadership shared with members and their families about behavioral health?
- Do your department members feel comfortable asking for help when they struggle?

STRATEGY STOP

Integrate behavioral health resources in the fire department and create a culture open to discussing areas of struggle.
In the past decade, the U.S. fire service has done a better job of openly talking about mental health issues and removing the stigma of seeking treatment. For example, the NVFC recently implemented the Share the Load support program for fire and emergency services (fig. 7–8). This program strives to

Figure 7–7. Informational flyer about the NVFC First Responder Helpline

educate firefighters on mental health issues and offers departments and their volunteers with a list of mental health resources.[32]

Share the Load Resources for Volunteer Fire Departments
Psychologically Healthy Fire Departments Initiative (NVFC)

- *Directory of Behavioral Health Professionals.* This is a vetted directory of behavioral health professionals with firsthand experience assisting fire and emergency services.
- *Psychologically Healthy Fire Departments: Implementation Toolkit.* This downloadable toolkit was designed to help fire service leaders promote and foster well-being among members.

Share the Load Helpletter
This newsletter contains articles from subject matter experts and fire service veterans offering tips, tools, and resources to assist first responders with taking a proactive approach in addressing behavioral health issues (fig. 7–9).

Figure 7–8. Logo for the Share the Load support program for fire and emergency services

Figure 7–9. NVFC flyer with mental health tips

Informational reports

- *Suicide in the Fire and Emergency Services: Adopting a Proactive Approach to Behavioral Health Awareness and Suicide Prevention*
- *Fire Department Assessment and Communication after a Firefighter or Emergency Medical Technician (EMT) Suicide*

Virtual courses free to NVFC members

- *Preventing and Coping with Suicide in the Fire and Emergency Services*
- *Behavioral Health for the Fire Service*
- *Addressing Substance Abuse, PTSD, and Other Concerns in the Fire Service*
- *Stress and Life Management—Finding Your Life Balance as a Volunteer*

STRATEGY STOP

Establish a fire service chaplain as an internal support.
The fire service chaplain is an individual who provides spiritual and other support to those in the fire service and those affected by emergencies and other crises. A fire service chaplain may be clergy or a layperson, male or female, firefighter or not. Fire service chaplains operate in a nondenominational fashion. In addition to aiding during a crisis, fire service chaplains can provide pastoral intervention and a link to community resources for members struggling with behavioral health needs.

Religious officiants of all different denominations are often willing to get involved with their local fire department when a crisis occurs. Trained fire service chaplains can bring people together to provide an effective response to crisis situations. Establishing an official fire service chaplain program using a religious officiant from the community or a member with a heart for service prior to a disaster or department need is key. Serving as a chaplain to first responders requires a different set of skills than those of a community-based pastor.

Fire Service Chaplain Training
Official training for those interested in serving as a fire service chaplain is offered by the Federation of Fire Chaplains (FFC). FFC offers fire service chaplains an outlet for both training and support. The mission of FFC is "to bring together persons interested in providing an effective Chaplain Service: to give aid, comfort and help to firefighters and their families; to work toward

the betterment of all areas of the fire and emergency service."[33] The FFC Training Institute for chaplains covers topics such as the following:

- Critical incident stress management
- Pastoral crisis intervention
- Grief
- Disaster chaplaincy
- Fire families: Family support and awareness program
- Fire department ethics
- Death notification
- Dealing with difficult deaths
- Comprehensive acute stress management
- Emotional resilience

Collaborate with Local Behavioral Health Partners

In today's culture of the first responder world it is imperative that career and volunteer fire organizations begin and continue the behavioral health education [of] its members and their families. This includes understanding PTSD, Moral Injury, addictions, depression, relationships, and even suicide awareness and prevention. One such asset is to provide culturally competent counselors and chaplains for the department. We can not only educate them about behavioral health but must provide several types of therapy resources from equine therapy to EMDR [eye movement desensitization and reprocessing], Brain Mapping, CBT [cognitive behavior therapy], and numerous others. Our goal should be for our members to have great careers but better retirements!

—Jeff Dill, founder, Firefighter Behavioral Health Alliance

It is important to identify and engage clinicians in advance of a crisis. Establishing and cultivating local partnerships with behavioral health professionals increases the foundation of support when services are needed. As volunteer fire departments implement behavioral health wellness programs, they should involve local partners at the development stage. Many behavioral health entities will welcome the opportunity to engage with the fire department.

Most counties have a local substance abuse and mental health board that can provide a directory of local providers. If you are unsure how to find that board in your county, use the Substance Abuse and Mental Health Services Administration website (SAMHSA.gov) to identify the state or county mental health authority in your state.

KNOW WHERE TO GO

As a former licensed chemical dependency counselor and social worker assistant, I cannot overly stress the importance of ensuring that your mental health resources and guidance come from educated and licensed provider. There are many individuals with a heart of service but who lack the formal training to deal with true mental illnesses. The title "crisis counselor" does not mean that the individual is licensed or has the extensive training to provide clinical treatment. Also, life coaching is also *not* therapy. Therapy is designed to help you heal. Only licensed therapists are qualified to provide mental health care. Check with your county or state mental health board to find licensed providers. The NVFC Helpline can also link you to a licensed provider within your community.

STRATEGY STOP

Follow these tips for building mental health resource partners to support your members.

- *Determine your department's needs.* What are you expecting to gain from the partnership? Are you looking only to create a local directory of behavioral health providers? Or do you want to create a list of providers who have experience specifically working with first responders? Do you want a local professional to serve as a subject matter expert as you develop a department program? Are you looking for guest speakers to speak to members or family members on behavioral health topics? Are you seeking literature on behavioral health topics and resources to place around the station?
- *Develop an outreach list.* Keep a database of those you reach out to. Include notes on the response received, services that could be beneficial, contact information, and any specific information that might be helpful when you implement your department's behavioral health

program. Learn as much as you can about the organization or person you are researching. Identify who the decision maker would be to enter a partnership.
- *Weed out the poor candidates.* Unfortunately, not all providers and organizations will be a good fit to work with first responders. Do not waste time fostering relationships with groups or individuals who will not help your organization meet your program's goals.
- *Create your pitch.* Once you have identified a list of potential partners, your goal is to schedule a face-to-face meeting to make your pitch for support. Include the following in your pitch:
  » A brief statement regarding the status of firefighter behavioral health in the United States, your goals to address first responder behavioral health needs at a department level, and what it will mean for the community you both serve.
  » Describe the benefit the partner will receive. For example, what recognition can you offer them in exchange for partnership? What impact will they be making on the community by serving first responders?
  » Provide specific information about your fire department. This is where the fact sheet outlined earlier in this book (see chapter 3) can be beneficial.
  » Present options for your partner's involvement (training, debriefings, counseling, resources, etc.). Be clear about what support you want from the partner. Be flexible—you might have to partner with multiple organizations and providers to meet your overall goals.
  » Finish the conversation by setting the next step. Who will contact who? When will the follow-up occur? Is there anything else that needs to be done?
  » Make the agreement, in writing. It is important that both partners understand exactly what is expected and when.
- *Keep an active relationship.* As in any relationship, be attentive to see it grow. Ensure that there is consistent communication with partners and updates. Share data and clarify current needs. Send notes of gratitude and find opportunities to show appreciation.

STRATEGY STOP

Create and maintain a community resource library online and in the station.
Create a digital and paper resource library where members and their families can access resources for behavioral health needs. Building a resource library might seem overwhelming, but it is simply collecting and categorizing resources. Ideally, offer a way to download materials from the Internet in the privacy of the members' homes, as well as maintaining a location for paper copies for members who are not comfortable with computers; paper copies will also serve as a visual reminder that behavioral health resources are available to members. Fire service leaders should frequently remind personnel and their family members that these resources are available and seek input on other resources that might be helpful.

Tips for Creating a Fire Department Resource Library

- Create lists of categories and subcategories for each topic you want to include or that your members have expressed interest in.
- Assemble helpful handouts and brochures from creditable sources for each topic.
- Document internally where handouts and brochures can be retrieved from, for refilling the library or updating outdated versions.
- Ask permission to scan and turn print resources into a Portable Document Format (PDF) file for a digital library, should there not be a digital version available.
- Monitor the digital resource library to ensure that links (i.e., uniform resource locators) to resources are still active.
- Partner with a local university to find an intern to help you develop your resource library.

Peer Support Programs

Encourage a culture of peer support. To address behavioral health challenges among volunteers, establish an organized *peer support program*, also referred to as *critical incident peer support*. A peer support program allows giving and receiving nonclinical assistance from peers who have had similar experiences. This peer support team should be deployed in the event of a line-of-duty death or injury, after a mass-casualty incident, and after difficult calls.

Peer support programs provide an avenue for firefighters to talk with trained peer supporters that can serve as a bridge to mental health professionals, offer educational support, and provide social support. Members of the peer-support program should be individuals who are highly respected and trusted. All peer-support members should be vetted and receive specialized training in providing peer support.

A peer support program in the volunteer fire service offers numerous benefits, as it provides a structured and empathetic support system for volunteers. Advantages include:

- *Emotional support.* A peer support program offers a safe space for volunteers to share their feelings, experiences, and challenges. It allows volunteers to process difficult emotions related to traumatic incidents, which can help prevent burnout.
- *Reduced stigma.* Encouraging peer support among volunteers reduces the stigma around seeking help. Volunteers are more likely to seek assistance from their peers, knowing that they will be understood and supported.
- *Improved mental health.* Peer support contributes to improved behavioral health outcomes for volunteers. This can help to prevent or manage behavioral health issues, such as stress, anxiety, and depression, by providing a network of understanding individuals.
- *Enhanced resilience.* Engaging in peer support activities promotes resilience among volunteers. This allows volunteers to learn coping strategies and gain strength from knowing they are not facing challenges alone.
- *Increased retention.* A strong peer support system enhances volunteer satisfaction and fosters retention. Volunteers who feel valued and connected have reduced likelihood of quiet quitting or early resignation.
- *Better team cohesion.* Peer support fosters a sense of camaraderie and team cohesion within the fire department. Volunteers engaged as a team become more supportive and understanding of each other, leading to a more cohesive team.
- *Early intervention.* Peer support programs can assist with detecting early signs of stress or emotional distress. Early detection can lead to timely interventions and support before issues escalate.
- *Professional development.* Peer support programs can offer training opportunities to volunteers. Additional training can enhance the skills of volunteers by encouraging in active listening, communication, and supporting others.

- *Improved performance.* Volunteers who feel emotionally supported are more likely to perform better. Performance is generally improved as a result of having a positive mental state and a strong support system.
- *Enhanced organizational culture.* Implementing a peer support program contributes to creating a positive and supportive culture within the fire department that values the well-being of all volunteers.
- *Community impact.* When volunteers receive the support they need, they can better serve the community and respond effectively to emergencies.
- *Family impact.* When volunteers receive the support they need, they can become better members of their families and peer circles. They are less likely to engage in negative coping skills as a result of unresolved behavioral health issues.
- *Volunteer satisfaction.* Knowing that they can rely on their peers for support and understanding enhances volunteer satisfaction, leading to a more committed and engaged volunteer force.

A peer support program is a valuable investment for fire departments. It fosters a culture of well-being within the department. Such programs can strengthen the department, enhance volunteer satisfaction, and ultimately improve the quality of service provided to external stakeholders.

STRATEGY STOP

Extend existing peer support programs to volunteers.
Boulder Crest is a nonprofit organization that offers programming at no cost to first responders and veterans. Dr. Richard Tedeschi is the executive director of the Boulder Crest Institute for Posttraumatic Growth. The concept of posttraumatic growth is the foundation of Boulder Crest's programs. Posttraumatic growth has been found effective for reducing symptoms related to PTSD, depression, anxiety, stress, insomnia, and negative affect.[34]

Boulder Crest has two key programs:

- *Warrior PATHH [Progressive and Alternative Training for Helping Heroes].* This is a 90-day, nonpharmacological, peer-delivered training program that begins with a 7-day intensive and immersive initiation at one of nine locations across the United States.
- *Struggle Well.* This program was built exclusively for first responders as a 5-day immersive program that speaks to the unique challenges, skills, and abilities of this remarkable community.

STRATEGY STOP

Form a peer support program for volunteers.
Starting a peer support team in a volunteer fire department requires careful planning and consideration. The following steps can assist with establishing an effective peer support team:

- *Identify the need.* Assess the current behavioral health and well-being support within the volunteer fire department. Identify any gaps or challenges that could be addressed through a peer support program.
- *Gain leadership support.* Obtain support and buy-in from the fire department's leadership. Leadership endorsement is crucial for the success and sustainability of the peer support team.
- *Form a core group.* Identify a group of dedicated and empathetic volunteers within the department who are willing to volunteer their time to be part of the peer support team.
- *Training and education.* Invest in your core group through training and education. Ensure that core members of your peers support program receive proper training in peer support, active listening, communication skills, confidentiality, and behavioral health first aid. Training will help your core team to effectively model and provide support to their fellow volunteers.
- *Develop policies and procedures.* Creating clear department policies and procedures for the peer support team is key. Outline the peer support team's purpose, their roles and responsibilities, confidentiality requirements, and the protocols for accessing external resources.
- *Confidentiality and boundaries.* Emphasize the importance of confidentiality and set clear boundaries for the peer support team members to maintain a safe and trusting environment. Communicate those boundaries within the department to ensure that all volunteers are aware of the protections in place to keep their interactions with the peer support team confidential.
- *Seek external resources.* Establish partnerships with external behavioral health professionals, counseling services, and local support organizations to provide additional resources or serve as a guidance board to the peer support team.
- *Promote awareness and acceptance.* Conduct awareness opportunities within the department to promote acceptance and understanding of the peer support program.

- *Create channels for communication.* Set up and document the communication channels (text, chat, email, phone line, etc.) for volunteers to reach out to the peer support team when they need assistance.
- *Establish regular meetings.* Schedule regular meetings for the peer support team to discuss ongoing barriers, best practices, and process improvement ideas and to plan future support initiatives within the department.
- *Promote self-care.* Encourage the peer support team and all volunteers to prioritize self-care. Fire service leaders should lead by example to promote a culture of wellness within the department.
- *Evaluate and improve.* Continuously evaluate the effectiveness of the peer support team. Seek feedback from volunteers (when they are willing to share) and make necessary process improvements to increase effectiveness.
- *Expand the team.* Allow others to join the peer support team to keep it from becoming cliquish and to ensure you have a pool of trained individuals.

Peer Support Resources

International Association of Fire Fighters (IAFF) Peer Support Training program
The IAFF offers a Peer Support Training program (https://www.iaff.org/peer-support/#about) although there is a cost. However, they also have a free 2-hour Behavioral Health Awareness Course that is self-paced and available to anyone, regardless of IAFF membership.

NVFC Psychologically Healthy Fire Department initiative
The National Volunteer Fire Council offers the Psychologically Healthy Fire Department initiative, which provides tools and resources to support responder health and well-being. The Psychologically Healthy Fire Departments: Implementation Toolkit is designed to help fire department leaders promote and foster well-being among their members. The toolkit covers six key categories:

- Member involvement
- Health and safety
- Member growth and development
- Work-life balance
- Member recognition
- Effective communication

Each category is examined along with specific actions that can be taken, special issues to consider, case studies from successful departments, and additional resources.

In addition to the Psychologically Healthy Fire Departments: Implementation Toolkit, the NVFC has developed supplemental training to help leaders create a fire department in which behavioral health is prioritized. A course is available in the NVFC Virtual Classroom (https://www.nvfc.org/phfd/#toolkit) and is currently free to everyone.

Critical Incident Stress Management Teams
Critical incident stress management (CISM) is another type of support program that has deep roots in the emergency services community, dating back to the 1970s. The National Interagency Fire Center indicates that the purpose of CISM "is to mitigate the impact of an event, accelerate the recovery process, and assesses the need for additional or alternative services."[35] CISM is a comprehensive, integrative, multicomponent crisis intervention system. CISM is viewed as a form of psychological first aid and should not be a substitute for psychotherapy. CISM is not a singular technique, and there are multiple components within CISM for use before, during, and after a crisis occurs.

The University of Maryland, Baltimore County Department of Emergency Services offers a certification in CISM (https://ccism-cert.org/) through its Professional and Continuing Education program. This is purported to be the world's first university-based CISM certification. They define CISM as "a specific corpus of knowledge relevant to the fields of psychological crisis intervention, sometimes referred to as 'early psychological intervention,' and disaster mental health."[36]

Provide Workplace Accommodations for Members in Need

As a fire service leader, you have an obligation to provide accommodations to volunteers in need. Behavioral health issues are not rare occurrences. In fact, the National Institute of Mental Health estimates that one in five people will experience a mental health condition in their lifetime and that employers most likely have at least one employee with a mental health condition.[37]

Most of us will experience a life event that distracts us from work. Common life events that can lead to a personal life crisis include divorce, death of a loved one, or an ailing family member. Some volunteers can separate home and volunteer life; others have situations that cause their personal and professional worlds to collide. It is important for fire service leaders to be empathetic while managing volunteers in these circumstances.

Know what accommodations you can offer. As a leader, listen first and don't pry. Simply ask the volunteer how the department can support them and help them to be successful. Some volunteer firefighters will need workplace accommodations owing to their behavioral health condition, and other volunteers with behavioral health conditions will not. For those volunteers who do need accommodations, they should be supported to ensure retention. The following are examples of accommodations, but these will vary per volunteer needs and should be individualized with input from the volunteer:

- *Leave of absence without repercussion.* Volunteers should be allowed to step away from volunteer duties to seek treatment for behavioral health conditions without the fear of being relieved of fire service duties, facing retaliation, or negative treatment from leaders and peers. If a volunteer takes a leave of absence, be sure to check in regularly with them.
- *Adjustment of membership requirements.* Volunteers may need to adjust their responsibilities at the fire department—for example, refraining from running calls and moving to an administrative role. Their membership requirements, such as attendance or call response, might need to be reduced.
- *Temporary change in duties.* A change in duties, such as serving as an officer, might need to temporarily occur. It is important that the volunteer needing the accommodation and the person(s) performing the duties during the change understand that the arrangement is temporary. Stripping an officer of a title because they are seeking treatment for behavioral health may do more harm than good.

STRATEGY STOP

Offer an Employee Assistance Program (EAP) to members. Some volunteer fire departments can use their locality's EAP. An EAP is a voluntary program offered by an employer that offers free and confidential assessments, short-term counseling, links to outside resources, and assistance with dealing with job-related problems. EAP addresses issues such as behavioral health, substance abuse, stress, grief, family issues, and psychological disorders. For departments that do not have a local EAP, the NVFC offers an EAP through the helpline that provides up to five counseling sessions for members at no cost.

STRATEGY STOP

Establish therapy dog partnerships for routine visits to the fire department.

Therapy dogs can provide comfort, affection, and love to firefighters. Many therapy dogs and their human handlers volunteer in clinical settings, such as hospitals, nursing homes, and schools. Establish relationships with therapy dogs in your community. Have them visit the fire department on a regular basis and ask if they would be willing to be available for debriefings after a difficult call. Importantly, to qualify as a therapy dog, the dog must be tested and certified for suitability. If you are unsure where to start, contact your local hospitals to ask if they have a therapy dog program and speak to the program coordinator.

Case Study: Mr. Riggs the Therapy Dog

There is limited research on the use of therapy dogs among the first-responder population, but the use of animal intervention to reduce stress has been shown to be effective.[38] In particular, dogs have been shown to reduce heart rate and reduce blood pressure.

In 2022 a tiny basset hound puppy joined our family with a purpose—to become a certified therapy dog to serve the community during fire prevention activities. Mr. Riggs, named after *Lethal Weapon*'s Martin Riggs, started training to become a therapy dog before he turned 1 year old (fig. 7–10). What I didn't expect, though, was for him to pick up on my own postcall and life anxieties.

As a former federal agent, I am always on alert for terroristic events. While in training, Mr. Riggs accompanied me to church. As we all stood to sing, I started to do a quick threat assessment of the room and plan for escape in my head. This is because I had been trained for years that houses of worships are a target, and that training does not leave you when you become a civilian. In fact, it makes what should be enjoyable moments stressful.

Mr. Riggs picked up on my anxiety as I was conducting this assessment in my head. He stood and nudged my knee. As soon as my attention broke from the intrusive thoughts, he laid back down. This illustrates what research

Figure 7–10. Picture of Mr. Riggs for a challenge coin

has shown: Therapy dogs reduce distress levels and increase comfort during stressful situations.[39]

Not every dog was born to be a therapy dog. Therapy dogs should undergo training. Although there is no national standard or certifying group, psychology and animal interest groups have stepped up to help. For example, Therapy Dogs International is one volunteer group that provides qualified handlers paired with trained therapy dogs for visitations to institutions, facilities, and any other place where therapy dogs are needed.[40] The AKC Therapy Dog title recognizes the following programs:[41]

- Alliance of Therapy Dogs (formerly Therapy Dogs, Inc.)
- Bright and Beautiful Therapy Dogs
- Love on a Leash
- Pet Partners (formerly Delta Society)
- Therapy Dogs International

Thoughts to Consider

- How can therapy dogs benefit first responders?
- How should you identify a dog who is appropriate to be a therapy dog?
- Is there a recognized training program in your area that you can reach out to?

Work-Life-Volunteer Balance: When the Scales Are Tipped, Disengagement Occurs

The challenge of balancing the needs of home life, work, and the volunteer organization is not unique to the fire service. Over 41 million people working and serving a volunteer organization indicated a struggle to balance home and the volunteer organization.[42] However, unlike other community volunteer opportunities, volunteering as a first responder is one of the most time-consuming volunteer commitments. *Work-life-volunteer balance* is a phrase I coined while conducting my research to describe the challenge volunteers face with trying to balance a career, a personal life, and volunteering in the fire service life. Volunteers who struggle to find a healthy balance—which is *not* an equal balance—are at risk for fatigue, burnout, and decline in productivity.

Work-life-volunteer balance challenges that volunteer firefighters face include working in a paid career while responding to emergency calls, as well as completing training hours to earn and maintain first-responder certifications, performing fire department administrative duties, and engaging in their family life, social life, and personal development (school and career training). When a firefighter is forced to pick between the fire service and family demands, resentment and disengagement can occur. In turn, the inability to balance both the fire service and family commitment can lead to burnout and turnover.

Many studies on work-life balance (notice no "volunteer") suggest time and emergency management as a strategy. The nature of the fire service itself with the unknown of what and when of an emergency leads to difficulty for a volunteer to plan ahead, delegate, and streamline activities to maximize time and energy. Several internal and external factors in combination contribute to a greater work-life-volunteer imbalance for volunteer firefighters. These factors are outlined in the following sections.

Dual-Income Families

A volunteer fire service family no longer looks like it did when our fathers and grandfathers started volunteering. Over time there has been a shift from single- to dual-income couples in the United States. The percentage of dual-income households has been climbing since the 1960s,[43] with 62.3% of families with children currently having both parents working.[44] In addition, 70% of volunteer firefighters are married or in a relationship.[45]

The shift to dual-income families leaves less time for people to volunteer because time is spent on their career, parenting, and maintaining the home. Thus, our firefighters are faced with trying to balance energy between home, work, and the fire department. They want to be a good partner, they want to be a good parent, they want to be a good employee, and they want to be a good volunteer.

U.S. employees in general are struggling to find a balance that keeps them healthy. Many are working overtime or second jobs to meet the basic needs of life. We are also at a place in time in which we have reverted to unhealthy and unsafe work practices that were in place a hundred years ago. There is a lack of healthy boundaries in the workplace, with employees being asked to do more with less compensation. This is the reason why in 1940 the Fair Labor Standards Act adopted a 40-hour week. Exempt salary employees in the United States are often taken advantage of, with employers expecting them to work more than 40 hours a week and not giving them the ability to self-determine how many hours they need to spend in the office. Gone are the days when a salary person could respond to a fire call without taking vacation or paid time off.

Training Demands

Fire service and EMS industry training demands have increased tremendously over time for the volunteer. In addition to the time demands faced by dual-income families, the increase in training requirements and standards poses a challenge for volunteers wanting to serve. When I first started with the Ohio volunteer fire service 2 decades ago, a volunteer firefighter could make entry during a structural fire with a 36-hour volunteer firefighter certification. Now that course is viewed as an "introductory, awareness-level course" only, and under this certification "prohibited activities include environments which are considered to be 'Immediately Dangerous to Life or Health' (IDLH), including but not limited to, hot zone operations at uncontrolled fires or hazardous materials releases involving fixed structures, mobile equipment, or outdoor areas as well as operation of emergency vehicle apparatus."[46] Ohio now requires a Firefighter 1 course of 160 hours as the minimum level of certification to provide firefighting services in the State of Ohio. This increase in hours is not limited to Ohio—this trend is happening across the country to ensure volunteers are qualified to perform firefighting duties.

Although the "36-hour card" was an option when I entered the fire service, my department required volunteers to have taken Firefighter 1. I was a single mom who worked full time while raising three school-aged children. I was left to find child care two nights a week and every weekend for fire school. In addition to fire school, I had to find child care for department training and meeting nights. This made for a very stressful and challenging few months.

My story is not unique: Many fathers and mothers have the same struggle in balancing life and the entry requirements to become a firefighter. Many are unable to juggle it all and drop out while going through the training to become a firefighter.

Firefighter training is not the only requirement included in the startup training to serve as a volunteer. Our industry expects many skills along with continued training. Many fire departments offer specialized services that require additional training and certifications that a volunteer must pursue while balancing work and life (table 7–1).

Training time requirements for becoming an active member can be a deterrent for those exploring the possibility of volunteering. Beyond the startup demands of the initial training, ongoing continuing education can also become hard for volunteers. Training demands are not only impacting new members but also seasoned members who are no longer able to run calls under the new certification standards; these members have been performing the job for years but are now no longer allowed to function until meeting an additional training demand.

Table 7–1. Percentage of registered fire departments providing specialized services[47]

Specialized Service	% of Departments Providing
Vehicle extrication	77.7
Fire/injury prevention/public education	64.0
Wildfire/wildland-urban interface	63.1
Basic life support	59.9
EMS—nontransport response	39.8
Technical/specialized rescue	35.8
Fire investigation/fire cause determination	35.4
Fire inspection/code enforcement	34.9
Departmental (in-house) training academy	22.0
Advanced life support	21.4
EMS—ambulance transport	21.1
Hazardous materials team	17.9
Juvenile fire setter intervention program	13.7
Airport/aviation	8.0
Fireboat	4.8

> **STRATEGY STOP**
>
> **Use online and hybrid training models to provide a balance for volunteers.**
> Training does not all have to be fully in person. Fire service leaders can adopt an online or hybrid training model and meet the same goals. The goal of training should be quality of engagement, not quantity of individuals. Fire departments can accommodate schedules by assigning online training modules or videos that can be reviewed outside of the department. To ensure member engagement, departments can create online discussion boards or group chats about assigned training that can include specific questions.
>
> Online training can be done as a prerequisite to a hands-on training. This reduces the time spent at the department and accomplishes the same goal. This means that firefighters do the classroom part of training by watching a video before being tested on the hands-on portion at the station.
>
> This does not mean that you reduce training standards or even stop offering training. For instance, you might still do an in-person training demonstration but record it and post the video on a password-protected YouTube channel; thus, the recording could be accessed online by volunteers who have other commitments, such as work or family. This ensures that everyone gets the *same* training regardless if at the station or learning from home—for example, while feeding a baby or taking care of an aging parent.

Increased Call Volume Owing to EMS

In 1973 my great-grandfather Dr. J. Fred Lembright was interviewed for a radio show about where emergency medicine was headed. The interviewer asked if he thought medicine was going in the direction of taking care of more people better or did he think that they were sliding backward? My grandfather, a man who made numerous house calls in his career, responded with two predictions:[48]

- First, he predicted that EMS would provide more "doctor" type of medical practice in the field, namely paramedic and advanced life support services: "We are at a crossroads. There are many so-called delivery types of care at least being tried experimentally in the country today. Some of them depend entirely upon medical doctors to give the services. In some, there is a tendency to use nonmedical but trained personnel to do a lot of the basic work, and opinions and judgment [are] only to be given by the medical personnel. In this respect—perhaps where too much emphasis is placed on nonmedical personnel—there may be a slippage in the kind of care that's given."

- Second, he predicted an increase in the use of emergency departments: "There are several factors that will increase the use of the emergency room. Number one, sometimes it is more convenient for a family or a patient to be seen in the emergency room because it works better for the family routine, particularly if both parents work and the mother doesn't get home until five or six o'clock in the evening. They feel that they can get quicker services in some cases, without having to wait so long as they may have had to wait in our offices of 25 years ago. There will be a number of people who will use the emergency room facilities because it comes under the insurance program which is carried for them by their employer or some other group. The feeling of being able to see one of several doctors that may be indicated will lead some people to go to the emergency room and also receive medication if necessary."

My grandfather died 2 years before I was born. I never got the opportunity to meet him, but I know that the emergency department staff loved him. I know this because when I became an EMT, the first time I walked through the ambulance entrance with a patient, there was a plaque on the wall dedicating the emergency department to him—and in fine print it said the plaque had been purchased by the emergency department nurses.

My grandfather predicted an increase in EMS because of shifting family dynamics, insurance, and drug seeking. This increase has significantly impacted the volunteer fire service. As mentioned in chapter 5, the NVFC has reported that the volume of calls volunteers are receiving has tripled, with medical calls representing the greatest proportion of this increase. Almost 75% of volunteer EMS professionals work in rural communities, compared with 30% of paid EMS.[49] Many seasoned volunteers remember a time before emergency medical calls were a part of the fire service. I am amazed that my grandfather could see where the United States was headed in emergency care even before EMS had become fully integrated within our society.

The increasing level of medical calls in the volunteer fire service has drastically increased the response requirements placed on volunteers. With many volunteer fire departments now transporting patients, the length of time spent while called out has expanded. Some attribute the increase in calls to the popularization of EMS by media and television shows. An analysis of EMS calls showed a peak in calls during the winter seasons (26%), Saturdays (16%), and during the daytime (39%) with the following breakdown of reasons for calling 9-1-1:[50]

- Unclear problem, 19% (the most frequent)
- Wounds, fractures, or minor injuries, 13%

- Chest pain or heart disease, 11%
- Accidents, 9%
- Intoxication, poisoning, or drug overdose, 8%
- Breathing difficulties, 7%

When fire departments recruit new volunteers, they tend to recruit for "firefighters" without making clear that the volunteer's commitment will often extend to medical calls. Is this a failure to communicate realistic expectations from the start?

Automatic Alarm Response

We are all familiar with the false-alarm calls that can make us complacent if we are not careful. Calls triggered by automatic alarm systems, especially in areas with commercial buildings, have contributed considerably to increased call volumes. Many of these alarms are caused by malfunctions, causing frustration and complacency among firefighters. The National Fire Protection Association reported that in 2012 U.S. fire departments responded to 2,238,000 false alarms, meaning that 1 in 12 calls a fire department responded to was a false alarm.[51] Furthermore, many home and business owners fail to address malfunctioning alarm systems, leading to repeated calls to the same address. These excessive requests for assistance for false alarms lead to volunteer burnout.

Work Commute Requirements

Another shift impacting volunteer fire services is the growing distance between individuals and their jobs in communities served by volunteer firefighters. Jobs in the Rust Belt have moved overseas, and these small communities have been left abandoned. Individuals are living in one area and commuting to work in another, which can lead to community disconnectedness. With jobs not being based in the community, employers are less likely to let volunteers leave to respond to calls. According to the U.S. Census Bureau, the average commute time to work in the United States is 27.6 minutes, with almost 10% of commuters reporting a commute of at least 1 hour, and almost 25% of the working population commutes outside of their home county.[52]

Firefighter and Family Guilt

The demands placed on fire departments are not the only thing on the rise. These demands have tipped the scales of work-life-volunteer balance even further. Volunteer firefighters trying to balance their roles will inevitably let someone down—either their fire department family or their loved ones at home—and as a result, personal guilt level has increased. A firefighter who misses their child's extracurricular activity or rushes away from dinner to attend to fire department duties may experience *parent guilt*; a firefighter who misses training to be at their child's event or a social outing may experience *firefighter guilt*. We have created a culture in which we make our firefighters pick between the fire service or being at Little Johnny's soccer game.

We as the fire service have not done a good job at providing flexibility to meet family needs. Many departments continue to hold training and meetings on the same evening of the week and at the same time. Therefore, firefighters must pick between outside life commitments and fire service commitments. We also have a history of giving members who are absent from training or have a low call response a difficult time when they eventually do show up.

Missing key family events to attend volunteer fire service duties may lead to resentment—from volunteers themselves and from their family members. This resentment can lead to a lack of support and marital frustration. Firefighters who lack family support are at a higher risk of turnover. Another stressor on the family is the concern about the risks associated with serving as a first responder. There is often an underlying fear of death or injury.

Failing to balance the fire service and family life can lead to emotional exhaustion. As the demands from the fire service increase, volunteers are placed at higher risk for negative psychological and physical impacts. The invasive hours the fire service requires leads to fatigue and stress. The unpredictability of emergencies interfering with home routines can lead to decreased volunteer satisfaction. Volunteers with a healthy work-life balance are more likely to experience higher levels of organizational commitment, goal achievement, better health and wellness, and a happier home life.[53]

Generational Factors: The Impact of Unmanaged Generational Differences

There is no alternate generation hanging out in a parallel universe waiting to swoop down and save your organization. They are it. They're already here—and they have many of the answers to the challenges we face in connecting with them. We need to get on with the business of training OUR replacements!

—Tiger Schmittendorf, fire service recruitment expert

Generational factors are a significant concern relating to firefighter retention. *Generations* are groups of individuals with similar ages and shared experiences. Fire service leaders face challenges with managing middle-aged, early career, and youth volunteers differently compared with previous generations.

Generational awareness is the understanding of how events during a certain period of time—such as political, economic, social and technological changes—lead to common traits among a group.[54] Unmanaged generational differences in the fire department can impact organizational commitment, job satisfaction, and retention outcomes. Generational conflict may exist between diverse generations of firefighters because of diverse perceptions of work-life balance and work ethics.

Generational Blame

Currently five generations are in the U.S. workforce: the mature generation, baby boomers, Generation X, Millennials, and Generation Z. Shared generational identity results in similar volunteer expectations in the form of psychological contracts. However, a violation of these psychological contracts can lead to turnover, lack of commitment, and a negative emotional response.

It is not uncommon for older firefighters to blame younger generations for the current volunteer firefighter crisis. However, the spirit of volunteerism in general is still alive among Americans under 35 years of age, of whom around 23% donate time to volunteer organizations.[55] It is critical for retention that fire service leaders gain a better understanding of how multigenerational employees can cohabitate within an organization and leverage one another's strengths.

Miscommunication and assumptions are a major issue when working in a blended generation. A blended generation is a multigenerational workforce in which different generations work together. This can include baby boomers, Generation X, millennials, and Generation Z. Fire service leaders face unique challenges when managing middle-aged, early career, and youth volunteers differently than they did with previous generations. It is important for leaders to recognize the factors that trigger disengagement in each generational group. Understanding these triggers will allow leaders to implement strategies to prevent turnover.

CASE STUDY: GENERATIONAL ASSUMPTIONS

I was working with a fire department to implement strategies for retention of volunteers. They had a high turnover rate and could not figure out why. To improve the culture, I took time to observe interactions between members and their leader during training nights.

One night, the EMS captain introduced a new Life Pack the department had just received. He called on a young medic student to come up and demonstrate how to use it. This young man slid over to the corner and pulled out his phone. The captain became irate and shouted to me that this was the disrespect he was talking about. What he failed to see was that the medic student was using his phone to Google how to use the new Life Pack. The captain had never gone over with him how the new equipment operated, so the young man did not want to be viewed as incompetent. When I pointed this out to the captain, he realized the young man was engaged and never intended to be disrespectful and indicated he would be more mindful to generational differences in learning in the future.

Another example is an instructor who was frustrated that he had to teach young firefighters how to change a tire. He was in disbelief that they were never taught this life skill. I had to remind him that while many of us grew up in an era without cellular phones and had to rely on ourselves when we got a flat or ran out of gas, younger generations have always been able to pull their mobile phone out to call for assistance. Hence, they never needed to learn how to change a tire the way many of us did. Why are we criticizing them for the advancement of technology?

Things to Consider

- How has technology changed since you joined the fire department?
- How can departments use technology-based learning tools?
- How can generational assumptions lead to poor working relationships?

Defining the Generations

The Mature Generation

The *mature generation* generally includes those born prior to the end of World War II in 1945. This generation is also referred to as the *traditionalists*. As a whole, this generation has less formal education and tends to be more conservative than later generations.

Individuals from the mature generation are said to have higher levels of commitment to organizations compared with members of other generations. Retention issues with the mature generation are often tied to age-related health issues that interfere with service. Individuals in the mature generation also start to disengage when the psychological contract changes.

Baby Boomers

Baby boomers include those born between 1945 and 1964. Within volunteer organizations, 66% of volunteers who return to a second year of service are from either the baby boomers or the mature generation. While there is a decline among the number of volunteer firefighters, the average age of U.S. volunteer firefighters is also increasing. Baby boomers have a higher level of comfort working with others and prefer teamwork compared with those in younger generations.

Inadequate leadership is one of the top reasons why baby boomers leave an organization. Organizational leaders also face the challenge of a volunteer gap when the Baby Boomers retire. Compared with the mature generation, baby boomers are more likely to maintain a volunteer role within an organization beyond the standard age of retirement.

Generation X

Generation X includes those born between 1965 and 1979. Generation Xers are more likely to have grown up with single parents balancing work and family or with parents both working and sharing child care. Individuals belonging to this group are more independent, competitive, and self-reliant than baby boomers. Compared with the baby boomers, Generation Xers have a greater preference for working by themselves over serving as a team.

Generation Xers also have lower levels of organizational commitment than do older generations. There is also a decrease in organizational loyalty among Generation Xers compared with older generations. Generation Xers place a high value on having flexibility with hours of service and work-life balance.

Millennials

Millennials include those born in 1980 or after. Another name for millennials is *Generation Y*. Turnover among millennials is higher than among older generations. Some organizations invest large amounts of resources to retain and engage millennials and are unsuccessful because of a lack of recognition for generational differences. Newer generations of volunteers often lack time or interest in volunteering.

The values of millennials are significantly different from those of the baby boomers and Generational Xers. Millennials place high value on work-life balance and are more apt to choose lifestyle over career decisions. Millennials have a higher turnover rate compared with other generations, with the average employed by seven different organizations over a span of 10 years.

The millennial generation not only looks different than older generations, but there is also a difference in thinking and behavior. Millennials tend be more self-centered, have higher self-esteem, and experience higher levels of anxiety and depression than older generations. Millennials place greater value on trust, dependability, dedication, and feedback.

The Newest Generation: Generation Z

Generation Z incudes those born after 1996. This generation has transitioned into adulthood during economic, political, and social turmoil. When I conducted my study, Generation Zers were just entering the workforce, so there is little data yet on retention rates. Most of this generation, which is projected to be the best-educated generation, entered the workforce during the pandemic.

The demographics of this generation are more racially and ethnically diverse, meaning the fire service must work harder to recruit volunteers who resemble the communities they serve. According to the Pew Research Center, 25% of Generation Z are Hispanic, 14% are black, 6% are Asian, and 5% are some other race or two or more races.[56]

Fire service leaders need to respect and find ways to engage Generation Z early on. Generation Zers are said to have a give-back attitude and are passionate about creating social change. Learning from the financial hardships of their parents and siblings, Generation Z is said to be more finically stable than those from older generations and start researching financial planning as early as 13 years of age.[57] Generation Zers have demonstrated a pattern of already giving their time and money to nonprofits.

Identify Diverse Ways to Manage Volunteers from Diverse Generations

There is a place for everyone in the fire service, regardless of age. Members from every generation can benefit the volunteer fire service, and it is up to the fire service leader to tap into the special skill set each generation brings and to implement strategies that address generation gaps. Managing volunteer firefighters from different generations requires understanding their unique characteristics, preferences, and communication styles.

STRATEGY STOP

Learn to manage multigenerational volunteers to prevent frustrations that can lead to disengagement.
The following strategies can help you to effectively manage a multigenerational volunteer firefighting team:

- *Promote department inclusivity and diversity.* Emphasize the value of diversity within the fire department and create an inclusive environment where all generations feel welcome and respected. Is your culture one that welcomes and includes all, or have you been catering preferentially to one generation?

- *Open the door for communication.* Encourage respectful, open, and transparent communication across generations. Fire service leaders should be willing to listen to all perspectives, feedback, and process improvement ideas. Do your volunteers know that they can talk to leadership and feel safe sharing?
- *Recognize individual strengths.* Acknowledge the strengths and expertise of volunteers from different generations. Each generation brings unique skills and experiences to the team that can be leveraged to meet the goals of the fire department. Embrace these different skill sets and make sure that each generation knows that what they bring to the table is of value. How does your department recognize individual gifts each volunteer brings?
- *Flexible training and learning.* Offer training and learning opportunities that cater to the different learning styles and preferences across generations. Some volunteers may prefer traditional classroom-style training, while others may prefer online or hands-on learning. How has your department changed your training to meet different learning styles?
- *Use technology wisely.* Use technology to streamline communication and administrative tasks, but ensure that this is done in a way that is accessible and user-friendly for across all generations. Are the tools used by your department ones that all volunteers are comfortable with?
- *Establish mentoring programs.* Facilitate mentorship opportunities between volunteers across generations. This fosters knowledge sharing and professional development, as well as building team relationships within the fire department. Have you established a mentoring program or activities to bridge the generational gap?
- *Provide flexibility in roles.* Allow volunteers to choose roles and responsibilities that align with their specific skill sets and interests. Ask them what they are interested in doing and what new skills they want to learn. Don't assume that they are content with the status quo. Flexibility allows each generation to contribute in the ways that suit them best. Have you surveyed your members to find out what skill sets they have and what they want to learn?
- *Respect work-life-volunteer balance.* Recognize that different generations may have varied priorities outside of the fire department. Be understanding and accommodating of work-life-volunteer balance needs. Avoid being critical of or teasing volunteers who choose to prioritize family over the fire service. Do your volunteers know that the department encourages family and personal success over department commitment?

- *Bridge the generation gap.* Organize team-building activities and social events to foster camaraderie and bridge the generation gap within the fire department. Choose activities that all members can participate in. Do the team-building activities you offer engage the whole team?
- *Give continuous feedback.* Provide regular feedback to volunteers across all generations, not just the new recruits. Recognize contributions and provide constructive guidance for improvement to all volunteers despite their department longevity. What type of feedback do you give to new versus seasoned members and to volunteers across generations?
- *Lead by example.* It is critical for fire service leaders to demonstrate positive leadership behaviors. Leadership should be proactive in promoting teamwork, a culture of respect, and cooperation across generations. What story do your actions tell about the value of inclusiveness in your department?
- *Adapt leadership styles.* Fire service leaders need to be flexible in their leadership approach. Some volunteers may prefer a more hands-on approach, while others may thrive with autonomy and self-guidance. Do you know how your volunteers want to be led?
- *Promote collaboration.* Fire service leaders should encourage cross-generational collaboration during training, projects, and recruitment and retention initiatives. This will allow your volunteers to learn from each other, foster diverse perspectives, and generate innovation. How have you engaged volunteers across generations to collaborate?
- *Resolve conflicts promptly.* Fire service leaders must address any conflicts or misunderstandings that may arise between volunteers from different generations. Unresolved conflict can lead to a greater team division. What process do you have in place to address conflict?

Additional Tips for Managing Multigenerational Volunteers

- The millennial generation thrives on immediate and frequent direct feedback on job performance; money is not the main motivator for millennials.
- Two key retention factors among millennials are performance incentives and opportunities for career advancements.
- Generation Xers view paid educational opportunities as an incentive. Hence, offering professional development to Generation Xers as a retention strategy can lead to strong replacements for the aging baby boomers.

- To balance the generational differences between millennials and older generations, fire departments can offer a two-way mentoring program.
- Generation Z values social interaction. Therefore, creating a culture of team spirit, camaraderie, and collaboration across generations will be important to Generation Zers.

STRATEGY STOP

Challenge stereotypes and educate members to prevent frustrations that can lead to disengagement.
Challenging stereotypes in the volunteer fire service is crucial to promote equal opportunities. There is a need for fire service leaders to challenge negative characteristics assigned to each group. It is important to educate fire department members about generational assumptions and the realities each generation has faced over time. Openly discuss how the era you were raised in impacts how you see and operate within the world.

The following strategies can assist with challenging stereotypes effectively:

- *Provide education and awareness.* Provide department-supported education and training on such topics as unconscious bias, to all volunteers and their family members. Increased awareness and education is key to challenging and dismantling stereotypes. Are you members aware of how personal bias impacts how they interact with others?
- *Lead by example.* Fire service leadership must lead by example to demonstrate an inclusive and respectful culture. This sets an example for all volunteers to follow and sends a message about what is and is not acceptable. Are you aware of your own biases and how they impact your interactions with others?
- *Promote diversity in leadership.* Encourage diversity in leadership roles to break traditional stereotypes. Note that this does not mean entitlement; rather, leadership is not limited to certain demographics. Fire service leaders can mentor those from underrepresented backgrounds to prepare them for future opportunities. How do you give everyone an opportunity to succeed within your fire department?
- *Communicate values.* Fire service leaders must clearly communicate the values of the department, emphasizing inclusivity, respect, and equal treatment for all volunteers. Emphasize a zero-tolerance policy

for treatment that does not reflect inclusiveness, respect, and equality. Are the values within your fire department openly known and discussed?
- *Celebrate diversity.* Importantly, fire departments should recognize and celebrate the contributions of volunteers from diverse backgrounds. This not only spotlights the strength of a diverse team but also gives role models to those who look like or come from the same background as those individuals. Do you recognize and celebrate the differences within your organization that make your members unique?
- *Challenge language and behaviors.* Fire service leaders should be proactive in addressing language and behavior that perpetuate stereotypes. Leadership should encourage a culture of open discussions to promote understanding and learning. Do you know how to address stereotypes in a way that fosters growth and not confrontation?
- *Practice inclusive recruitment.* Adopt inclusive recruitment practices that reach out to diverse communities and ensure equal access to volunteer opportunities within the fire department. Work with local leaders in the community to develop inclusive recruitment strategies. How have you engaged with community leaders to reach underserved and previously unreached areas?
- *Have diverse public representation.* Represent your fire department in a diverse and inclusive manner during public events, in print and digital materials, and in community outreach programs. Do the marketing materials you distribute look like community you serve?
- *Establish mentoring programs.* Mentoring programs should pair volunteers from different backgrounds, generations, and experiences. This will foster mutual learning and understanding. Do you have a formal mentoring program in place for growth across demographics and generations?
- *Encourage feedback.* Fire service leaders should encourage volunteers to provide feedback and share their experiences to identify areas where stereotypes persist within the fire department. Do your volunteers feel safe and comfortable speaking up?
- *Collaborate with other organizations.* Departments can partner with organizations that offer training in and promote equal opportunity efforts. They can share best practices and work together in challenging stereotypes. Are there organizations within your community that can provide education to your members?
- *Perform regular evaluation.* Fire service leaders should regularly evaluate their department's progress in challenging stereotypes and

creating an inclusive environment. How often do evaluate your progress, and what benchmarks do you use to monitor progress?
- *Publicize success stories.* Departments should share, with permission, the success stories of volunteers who have challenged stereotypes and overcome barriers within the volunteer fire service. How often do you share department wins on social media or with local news outlets?

Generational Learning Opportunities

- NVFC offers a four-part online series on Training the Next Generation, in the NVFC Virtual Classroom[58]
- The Federal Emergency Management Agency (FEMA) has published a downloadable document titled, "Generational Perspectives in Emergency Management—a Glimpse into Understanding"[59]
- The Office of Personnel Management has published a downloadable document titled "Leading a Generationally Diverse Workforce"[60]

STRATEGY STOP

Implement a two-way mentoring program to bridge the generational gap.
Fire service leaders can implement mentoring programs to reduce turnover tied to generational barriers. Mentoring and apprenticeship programs are important for both the recruitment and retention of volunteers, by increasing on-the-job confidence and performance.

Mentoring yields a positive on retention for both younger and older generations. To balance the generational differences between millennials and the older generations in the fire department, offer a *two-way mentoring* program. A two-way mentoring program works best when organizational leadership recognizes the millennials' or Generation Zers' expertise with technology and pairs them with a seasoned volunteer. The older volunteer provides shelter to the newcomer and navigates them toward success; the millennial or Generation Z volunteer benefits the older volunteer, who may struggle to embrace technology tools in the workplace, by transferring knowledge about and expertise with newer technologies. This pairing of younger and older generations can benefit both the volunteers and the fire department, by building confidence, comfort, and skills across age groups.

Tips for Starting a Mentoring Program
(Adapted from the U.S. Patent and Trademark Office[61])

Identify a program coordinator and an officer-level champion:

- Determine who will serve as the program coordinator to oversee the development, implementation, and monitoring of the mentor program. (This can be a non–first-responder volunteer.)
- Provide training and resources for the program coordinator to accomplish the assigned responsibilities.
- Identify an officer who will serve as champion to communicate the purpose and need for the mentoring program to internal and external stakeholders.

Identify the purpose of the mentoring program:

- Determine what you want the mentoring program to accomplish.
- Determine what you want the mentees to know or be able to do when they complete the program.

Identify who the target mentors and mentees will be:

- Identify and document the requirements to serve as a mentor.
- Identify and document the expectations for serving as a mentor.
- Determine who will be served (all members or just new recruits).

Establish the program approval process:

- Identify who needs to approve program components.
- Establish and document the steps of the approval process.

Establish policies, procedures, and responsibilities:

- Determine the length of the mentoring.
- Determine if mentees will rotate between mentors and, if so, the frequency of rotation.
- Establish the roles and responsibilities for the officer-level champion, program manager, mentors, mentees, and stakeholders.

Schedule activities to support mentors and mentees:

- Train and educate mentors on their responsibilities.
- Provide training on working across generations to both mentors and mentees.
- Identify activities for socialization and team building for both mentors and mentees.

Create program documents and resources:

- Develop a mentor application form.
- Create mentor and mentee agreement forms.
- Create activity logs.
- Establish a resource library to assist mentors (mentor guidelines, recommended training, and reading material).

Develop a communications strategy:

- Determine how the mentoring program will be promoted and how information and updates will be delivered (in person, by email, by text, in print, etc.).
- Identify how progress and success of the mentoring program will be reported.

Train and educate the mentor and mentee pool on aspects and benefits of the mentoring program:

- Identify how training will be delivered.
- Implement and provide ongoing mentor training.

Evaluate the program:

- Plan how you will evaluate the effectiveness of the program.
- Conduct evaluations with the mentor and the mentee about the effectiveness of the program.

STRATEGY STOP

Respect the time of volunteers in environments you can control.
The fire service has historically been the biggest offender of "time suck" among volunteers, and many fire service leaders have failed to implement strategies to respect time. The younger generations value time and desire a hassle-free environment to serve in. Fire departments that do not respect that will lose this generation.

Post a training schedule in advance—and abide by it. Start and end at the stated times. Chances are that the younger population has other commitments scheduled afterward, such as family or social plans, educational requirements, or work responsibilities. Older generations may have commitments to care for an aging parent or grandchild. Complicated and time-consuming systems will drive volunteers of any generation away. Respecting the time of volunteers sends a message that the organization respects the volunteer as an individual.

Tips for Respecting Volunteer Time

- Post start and end times of all meetings and trainings. This allows volunteers to balance outside commitments around fire service duties.
- Start and end meetings and trainings at the posted time. This ensures volunteers are not faced with an unplanned commitment conflict.
- At the end of a training or meeting, let volunteers know that the event is over and they are free to leave. This gives permission for volunteers to leave, even if the other members choose to stay and share stories.
- If a training or meeting runs over a posted end time, give members the freedom to leave without consequence or ridicule. This allows volunteers to meet outside commitments (child care, school, etc.) without guilt.

Organizational Climate and Culture Impacts on Retention and Recruitment

There is always knowledge we should acquire from one another.

—Bill Webb, CFSI

It is important for fire service leaders to be aware of how their organization is perceived and the values within it. These are your organizational climate and organizational culture, respectively. Organizational climate and culture are important for fire service leaders seeking to improve volunteer retention.

Organizational climate comprises the perceptions people attach to experiences that they have at work tied to decision-making, leadership, and work norms.[62] In other words, your organizational climate is how your stakeholders view the work environment. By contrast, *organizational culture* comprises the basic assumptions about the world and values guiding an organization.[63] Both culture and climate can impact volunteer behavior, attitude, and engagement.

Organizations with toxic culture and climates can experience higher rates of turnover, reduced quality of work, and increased liabilities. Such toxic environments can also negatively affect the services first responders deliver to community members. Volunteers who lack satisfaction with the organizational climate are likely to disengage and seek volunteer opportunities elsewhere. In particular, new members of the fire service are at great risk of turnover, with the highest risk during the first 3 to 6 months.[64]

It is essential for fire service organizations to create a culture and climate that fosters civil and respectful behaviors. Creating such an environment will lead to greater organizational commitment. An environment that allows bullying and other negative behaviors among firefighters will lead to high turnover. Poor leadership can impact levels of job satisfaction and retention among volunteer firefighters. Volunteers who find the environment or work unsatisfying are at greater risk of leaving.

Embracing New Volunteers by Organizational Socialization

Organizational socialization in the fire service is the process in which the new fire service members become acclimated to the culture of the fire department. It is important for fire service leaders to embrace new volunteers and foster social acceptance. Fire service leaders must provide opportunities for socialization between new volunteers and existing ones.

Organizational socialization should begin during the application process. The potential recruit forms an idea of what volunteering with brothers and sisters in the fire service will be like. It is important during the interview process and onboarding to familiarize the potential recruit with the organization's norms and values to determine if the relationship will be a good fit.

TIPS FOR FOSTERING ORGANIZATIONAL SOCIALIZATION

- *Communicate in real-time and not just through emails and memos.* Have actual conversations with volunteers and make them feel listened to. Sit down and eat with volunteers.
- *Offer an orientation for all new volunteers.* As part of the onboarding process, offer an orientation designed by current volunteers to help new recruits integrate into the department culture.
- *Utilize group chat features in your dispatch application features.* This helps to initiate conversations and connect with volunteers. This also allows volunteers to form relationships outside of the formal processes.
- *Plan group activities outside of the fire department.* The ultimate purpose is to bring volunteers together to bond whether at meals, charity work, sporting events, picnics, or something else.

The Influence of Job Satisfaction on Volunteer Retention

Creating an organizational climate that promotes job satisfaction among volunteer firefighters is another important factor for retention. Organizations that allow volunteers to have a fair amount of control over how to perform job duties are more likely to have satisfied volunteers. Creating positive work environments can also increase job satisfaction.

One study revealed that volunteer firefighters express lower levels of job satisfaction than career firefighters do, with only 38.5% of volunteer firefighters surveyed reporting high levels of job satisfaction.[65] The job duties of a volunteer can affect retention. If the work is unsatisfying, volunteers are at a greater risk of leaving the organization for a better opportunity. Older generations feeling ineffective in meeting organizational goals are more likely to leave the fire department. Importantly, public-sector employees have a lower level of job satisfaction than those in nonprofit or private sectors. This may be a result of frequent criticism from media, elected officials, and community groups or industry. The NVFC has cited poor leadership and management styles as hindering job satisfaction and retention among volunteer firefighters.[66]

STRATEGY STOP

Foster opportunities to eat together as a department.
"Kitchen table talk" is a phrase we often hear in the fire service. Meeting in an informal setting has many benefits among internal stakeholders. An environment where volunteers cooperatively eat together at the kitchen table can increase team unity and satisfaction. In particular, eating together fosters increased communication among team members who might not normally talk. While eating together is a common trait among career firefighters, volunteer organizations can also plan shared meals. Family members can also be encouraged to join in meals from time to time. Cookouts, potlucks, and community-sharing meals are cost-effective ways for volunteers to come together for a meal with minimal cost to the department. How do you bring your members together at the kitchen table?

Tips for Planning Potlucks at the Firehouse

- *Seek permission from department leadership.* This is vital to avoid scheduling a potluck during a conflicting event (training, community event, etc.).
- *Determine needs in advance.* Before posting the potluck invite, find out what the department will provide and what members will need to bring. Also determine who is invited—members only, or members and family?
- *Identify a date and time.* To minimize conflicts, look at training and event calendars.
- *Send invites through multiple channels.* Use email and messaging apps, post it in the department, and share it verbally. The invite will have more meaning if it comes from the chief.
- *Post a potluck sign-up sheet.* Use this to track what individuals are bringing and how many are coming. This can be posted as shared Google form or as a paper sheet in the station.
- *Post another sign-up sheet for volunteer needs.* Use this to identify volunteers willing to help set up or tear down. Again, this can be a Google form or a paper sheet in the station.

- *List any instructions that members might need to know.* Is the potluck a themed event? What time do they need to have their contribution there and how will it be stored to ensure food safety? Do participants need to bring serving utensils, their own beverages, chairs, or anything else?

Review Questions

1. Short sleep duration is defined as getting less than _____ hours of sleep within a typical 24-hour period.

2. The Help Network of Northeast Ohio recommends this easy-to-remember mnemonic to remember the signs of suicide: _____.

3. Generation Z are said to have a _____ attitude and are passionate about creating social change.

Discussion Questions

1. Why is it important for department leaders to understand the barriers to recruitment and retention?

2. How can you implement behavioral health resources within your department?

3. How can generational differences be leveraged as a strength?

8

BECOMING A CATALYST FOR CHANGE IN A WORLD OF RESISTANCE

The only person who likes change is a baby with a wet diaper.

—Mark Twain

The fire service historically has rebelled against change. Change in the fire service is inevitable. Seeing a chapter on change, some fire service leaders might have the urge to skip this chapter. They could be thinking, What does change have to do with solving my recruitment and retention issues? The answer is that if you want more recruits and better retention rates, you must be willing to change the way you are doing things. *You* must change, because society has already changed in many ways:

- Volunteers now expect more from the organizations they serve.
- The working population is opting for work-from-home roles, with work-life balance being a priority.
- Nonwage benefits are more attractive than compensation among volunteer firefighters.
- A positive environment is more important than a cool hat and title, as there are many places to volunteer within a community.
- Self-scheduling in the workplace is highly desired yet underutilized.

Navigating Change without Getting Lost

Change is difficult because it is tied to uncertainty. Uncertainty is often tied to anxiety and a feeling of lost control. What if building an organization that attracted talented firefighters and retained quality people was as simple as following a blueprint or map? This book *is* your road map to change. This book was developed to help you navigate change without getting lost, *but* you must be willing to do the work it takes to create change.

Leading an organization through change takes strategic planning and commitment from organizational leadership. I won't lie and tell you that it will be easy; to the contrary, managing change can be complicated and chaotic. Change is one of the most resisted things in the workplace and in life. It triggers emotions that a good leader needs to recognize and combat.

I once heard that change is like a rollercoaster—scary and exciting all at the same time. As you move forward on the ride, you bang from side to side, you often end up twisting upside down, and your emotions range from euphoria to panic. No matter what distractions occur on the recruitment and retention ride, the fire service leader needs to have the skill set to keep the team focused on the mission.

Importantly, change is just a label, applied to everything that we plan to do differently in the future. People dislike change because they prefer staying with what is comfortable, what is known. Staying with the known provides a sense of security. Change threatens stability.

Change can also mean starting over, back at ground zero. Starting over at the beginning can lead to feelings of powerlessness. No one likes to feel powerless. A sense of powerlessness can quickly turn to disengagement.

Leading the team through the change process in a positive method is critical during the implementation of the strategies you will learn in this book. Some will see the change as unnecessary, as they continue to blame others for the lack of volunteers. They will do nothing—and nothing within their organization will change. Others will see the need to change as a message that their past efforts were not valued, instead of seeing the past as a foundation for the future. What they should see is that it is their job to develop a future fire department that people will want to call their own and be proud to embrace.

Being a catalyst for change will require you to ensure that all team members have knowledge about the strategies and tools you plan to use for change. Lack of knowledge may result in lack of confidence, and this may ultimately derail implementation. All team members must know how to use the new methods effectively to prevent them from reverting to old ways.

CASE STUDY:
THE DAY PAC-MAN DIED

When teaching recruitment and retention in person, I often share this story about the importance of making sure *all* members are comfortable with the new tools being presented. A few years ago, my teenage son joined me on a road trip to speak at conference two states over. While at a rest stop, we stumbled across a vintage *Pac-Man* arcade game. I had not played the game in over 20 years but was excited to introduce my son to my childhood love. I slipped the quarter in, and all my expertise from the '80s came rushing back. I felt like a master on the first play as I beat each level—cherry, strawberry, orange, and so on. My son informed me that we could buy *Pac-Man* for the Nintendo Switch and play at home.

With my confidence soaring, I purchased the game ready to crush my family members. It wasn't long after we hooked it up that "the change" knocked my self-esteem right out from under me: Suddenly, I was powerless! As I tried to work the unfamiliar controllers, the ghosts gobbled me up. The screen was the same, the characters were the same, but the tool I had to use to achieve the goal had changed slightly. I became frustrated and lacked the joy I had had for the game when we were playing the arcade version.

Our seasoned firefighters—from the old school—often feel the same way about learning new tools. They might even try to backslide, like I did with *Pac-Man*. You see, I had what I thought was the solution: I could revert to old-school tools by buying a console-table version of *Pac-Man* to have at home—one with a joystick. I purchased the table, but there was a lack of interest from my family members. My son questioned why we would want to sit on small, uncomfortable stools to play on a small screen. He informed me that with the new way, we could all play the same game together while sitting comfortably on the couch and look at a larger screen. At that moment I realized that I had failed to adapt to the new tools of the game. I let my frustration with change get in the way of my willingness to learn a new tool that had a greater benefit (appeal) to the group (my family).

My frustration with change came from being pushed down to the beginning level in an environment that I felt I had been fully successful in before the change. I had lost control, powerless against the ghosts. I wanted to show my confidence by going back to the old tools. Our volunteers feel similarly frustrated when we introduce change without having a plan to help them navigate and learn. They get gobbled up, and as the players, they backslide to what is comfortable.

Things to Consider

- How can we introduce new tools without making our seasoned members feel like beginners?
- How can we get stakeholders to see that change can have a greater benefit to the group?
- How do we keep individuals from backsliding to what is comfortable?

The Reasons for Change

Change within the fire service is needed for survival. Most change is a response to a shifting environment, internal or external. Change may be forced, or it can be voluntary. It is key for organizations to change to meet the needs of both internal and external stakeholders. As communities, economic factors, and technologies change, fire service organizations need to develop strategies to keep up with these changing environments to avoid turnover and become attractive for recruitment. A good leader understands the need to help their employees through the change process to prevent employee derailment.

There are typically four reasons why an organization needs to implement change. These are structure, tasks, technology, and people. We'll look at each of these drivers of change in the next sections.

Structural Change

Structural change involves mergers, changes in hierarchy, management systems, downsizing, and restructuring. Structural change often comes with a high level of resistance, as those involved may view the change as threatening. Leadership must help volunteers to view the change as an opportunity to leverage resources to grow. When volunteers fail to understand the need for structural change, this can lead to turnover.

For example, a fire department making a shift from a paid-on-call volunteer department to a combination department would be a structural change. As part of this change, the need for volunteers to respond during staffed hours may be eliminated. Both the paid and volunteer firefighters would be reassigned to new positions. New policies and procedures would need to be developed to meet the structural change. This type of structural change would need to be strategically implemented to gain stakeholder buy-in.

Task Change

Task change entails responding to a need to change a process to improve quality, meet customer needs, and increase productivity. This often involves changes in job assignments or position, which becomes scary for volunteers. To minimize resistance, leadership must help volunteers see that a task change is an opportunity for improvement.

An example of task change would be a volunteer fire department making a change from having all vehicles respond to an incident to a new process in which only a select number initially respond. This task change would be a process improvement to reduce fuel costs and to ensure that all resources are not in one location, thereby allowing quicker response if there is a second incident. The reason the change is being made must be clearly communicated in advance of the change to ensure that members understand the benefit and the implementation process. Failure to do so may lead to resistance.

Technology Change

Technology change comes from innovation and impacts every organization. As technology evolves, the world changes. Much of the volunteer fire service has struggled over the years to transform with technology changes. Technology integrations require that volunteers learn a new skill set, which can be perceived as threatening. Hence, even while technology can lead to increased efficiencies, improved communications, and cost reductions, it can also lead to resistance. Introducing new technologies requires proper training and time for adjustment.

An example of technology change in the fire service is the significant change in the dispatching of calls. First responders not only get the alerts on mobile devices but also can verify their response since their location is being tracked. The use of electronic charting and mobile transmissions has also increased efficiency in volunteer emergency medical services, with providers transmitting vital supplies to the hospital before the patient even arrives. Patient care has increased as well, since doctors can see what they are getting into on a screen before the patient arrives. However, some older generations of firefighters have struggled to embrace using tablet devices on the scene.

People Change

People change not only results from a shift in consumers, volunteer turnover, and leadership change but can also stem from a change in a person's skill set, relationships, knowledge base, and attitude. A leader's attitude can set the tone of the organization. A leader is in unique position to help employees navigate the change process with a positive attitude. People change not handled properly can lead to disengagement.

For example, a volunteer fire department's chief steps down, and a new chief from the outside is appointed. The remaining officers do not approve of the chief who was brought in from the outside to lead their organization. The officers fall into the victim role, and their negative energy flows down the ranks. The new chief is met with resistance from day one and spends a great deal of time overcoming this challenge. The attitude of the officers robbed the department of the opportunity to come together during the change process to focus on building the future as a team. The authority appointing the new leader failed to obtain key stakeholder buy-in during the change process.

The impact that change has on volunteers will be based on how they view the direct impact to them. It is the leader's role to motivate the team to embrace change in a way that will allow successful implementation.

STRATEGY STOP

Provide a good reason for the change.
Fire service leaders need to give their team the *why*—the reason behind the change. Ensure that this reason reflects the values, goals, and mission of the organization. Change should occur only if it will positively support the organization's purpose. Thus, leaders must be transparent by communicating the why.

Transparency equals trust. In turn, disseminating factual information among stakeholders prevents rumors. Knowledge sharing should be a priority. Every volunteer in the organization at every level makes decisions and shares information every single day that has an impact on the future of the organization. Every decision made at every level counts. The more informed and aligned your volunteers are, the better will be the outcomes of those decisions. Don't be an information hoarder.

For example, if budget restraints are driving the change, then that should be communicated. Provide the facts; don't sugarcoat the story. A relatively simple change without reason can be more devastating than a difficult change based on a valid reason that is made clear. If there is not a good reason for the change, a leader should rethink the investment of time and resources that would be required to implement that change.

STRATEGY STOP

Be a change agent to motivate your followers.
A leader's attitude and response to change will set the organizational mood. Share with volunteers your commitment to seeing the change be effective. Communicate both your commitment to supporting the team during the process and why the change is personally important to you.

Allow volunteers the opportunity to share their feelings, to work through initial negativity and gain an understanding of the need for change. It is important to listen and ensure that the team knows that you are listening to feedback. Leaders may need to help "unfreeze" any volunteers who are resistant to change by exploring what fears are holding them back.

STRATEGY STOP

Be strategic.
A fire service leader needs to carefully think out the change process. A plan should be developed to engage those who will be affected in the change process. Avoid rapid changes, go slow, and allow people time to adjust during the process. Provide opportunities to learn any new skills needed for the change to be successfully implemented. Ensure the tasks and roles are clear to all volunteers to avoid any misunderstandings or rumors.

John Kotter offered the following eight steps for strategic change implementation:

- *Step 1: Create urgency.* Have open and honest conversation about potential threats, opportunities, and needed support.
- *Step 2: Form a powerful coalition.* Assemble the true leaders within your organization to influence others that change is necessary.
- *Step 3: Create a vision for change.* Write this out clearly in a way that captures how you see the future of the organization.
- *Step 4: Communicate the vision.* Demonstrate "walking the talk" by applying your vision to all aspects of your actions and operations.
- *Step 5: Remove obstacles.* This includes people and processes that are resisting, by implementing a structure for change that helps bring others in line with the vision.
- *Step 6: Create short-term wins.* Celebrate small wins in the journey and reward people who help you to meet the target wins.
- *Step 7: Build on the change.* Analyze after every win what went right and what needs improving, then build on the momentum that each win creates.
- *Step 8: Anchor the changes in corporate culture.* Talk about progress often, in particular by telling success stories, including change ideals during the orientation of new employees, and publicly recognizing key members of your change coalition.[1]

Communication Is Critical During Change

It is essential for fire service leaders to broadcast the change using effective communication strategies. You should have a two-way communication plan, to ensure that everyone who is impacted is kept informed and allowed to provide feedback. Offer formal methods for volunteers to communicate questions, concerns, and praise about the change process. Your communication plan should meet the needs of all generations and how individuals like to receive information. Survey your members and ask for their preferred method of communication. Given the variations in generational preferences, you may need to develop different avenues of communication to meet the needs of your internal stakeholders.

Importantly, the person facilitating the conversation regarding the change must create a safe environment for open discussion. The facilitator should encourage questions, feedback, and dialogue about concerns. The facilitator should be viewed as someone who listens respectfully without prejudice. As members take risks during the conversation, the facilitator should acknowledge those—and indeed appreciate them. Make sure your members understand that no repercussions will occur for opinions expressed during open discussion. If the facilitator does not have an answer to questions being asked, it is OK to say so and get back to the group with that answer later.

To keep your volunteers informed, I recommend using a team management tool or app. These tools allow team collaboration and transparency that can be centralized, with the entire team being notified at the same time. The fire service consists of nondesk, frontline employees. Sharing department updates with a mobile team management app ensures direct access and fast connection to wherever members are, since they won't be sitting at a work computer waiting on company emails. In addition to pushing out information, team management apps foster social interactions through forums for comments and questions. This provides an additional avenue for dialogue between leadership and members.

STRATEGY STOP

Foster team interaction through technology by implementing app-based technology (table 8–1).

Seek the Assistance of a Trusted Peer to Champion Change

Department leaders are not always the best people to champion change. One of the simplest ways to get buy-in to change is to assign a respected person to lead or co-lead the change effort. This should be an individual who will support the change and is respected throughout the ranks, to serve as its champion throughout implementation. The champion can then rally other volunteers to become trusted allies in the change process.

Table 8-1. Various apps can be used to facilitate communication among teams. Note that the application selected should be set up with the chief or another trusted officer as the main administrator to ensure continuity in the event of turnover.

Tool	Website	Notes
Crew	https://www.crewapp.com	Free for organizations with 35 or fewer members
GroupMe	https://groupme.com/en-US/	You don't need the app to use GroupMe. You can chat with your groups directly over SMS.
WhatsApp	https://www.whatsapp.com	Works internationally
IamResponding (IaR)	https://www.iamresponding.com	If your organization is using IaR for dispatching, you can tap into the team communication feature.
Active911	https://www.active911.com	If your organization is using Active911 for dispatching, you create an internal chat room for your whole agency, as well as smaller groups, such as stations or teams.

Typically, when volunteers feel trusted and empowered in the change process, they step up to lead. Moving an organization in a positive direction matters not only to the future of the department but also to those doing the work. Trust your members to do the work and support them as they help guide the process. Find opportunities for volunteers to get involved and use your change champion to make personal invitations.

Engaged volunteers will recruit others to be part of the process. Invested members aware of the problems the organization is facing will welcome an invitation to be part of the solution process. As the change agent, it will be your job to make them see why they are individually needed to achieve the goals. Make them see the value they can add to the organization. The more members you can rally to commit to the process, the greater the probability of success. Note that this does not mean that everyone will be satisfied with all decisions being made, but by engaging them, you will reduce misunderstandings.

At times personal loyalty to those resisting the change might supersede the commitment to moving the department forward. If we are not careful, this type of influence can derail change efforts. Meet resisters one-on-one to gain an understanding about why they disagree. Listen to their reasoning and let them know you hear their feelings. Try to identify what conscious or subconscious influencers might be affecting their resistance. Provide any additional information that might clarify the need for change or the process. Ask if they have an alternative solution to the change that is being suggested and ask them to share the positive and negatives of the change. Identify any additional training or skills that might be needed to remove any barriers.

Plan and Celebrate Short-Term Wins

Transforming an organization is not something that occurs quickly—it takes time. Fire service leaders should plan for and celebrate short-term wins to maintain the change momentum among stakeholders. The celebration of short-term wins can recharge those volunteers who are losing focus and commitment for the desired outcome.

STRATEGY STOP

Provide recognition of all-star change agents.
Be sure to take the time to recognize those who embrace the change process. Recognition is a motivator for increasing job satisfaction among volunteers. Be aware of and acknowledge any sacrifices made by volunteers. Rewarding volunteers for their efforts increases their willingness to take an active role in the change process and fosters a shared sense of responsibility among volunteers.

STRATEGY STOP

Be patient with the process.
Change does not occur instantly. It takes time for the seed of change to grow into something that can be harvested. If change occurs too fast, it can have the opposite of the intended effect: It can be destructive to the organization. With any change, a leader needs to take time to navigate the team through any pains tied to personal and political factors.

Resistance can occur when the implemented change is not viewed as having a 100% benefit for all stakeholders. If there will be a negative impact to an individual or a small group, give people time to adjust. Keep them informed during the process, be available for questions, and ask them to develop a better solution to solve the issue that doesn't negatively impact anyone. Communicate how much more will be gained over the small loss. Be sure all the stakeholders understand the implication for failing to change and be patient as they work through the process.

Review Questions

1. Leading an organization through change takes _____ and _____ from organizational leadership.

2. What are the four reasons an organization would need to implement change?

3. You should have a _____ to ensure that everyone who is impacted is kept informed and allowed to provide feedback.

Discussion Questions

1. Why is it important for fire service leaders to embrace change?

2. What methods does your department currently use for communication? How can you improve?

3. What are some methods to help your stakeholders through the change process?

9

DEVELOPING A RECRUITMENT AND RETENTION GAME PLAN

Leadership is about going somewhere. If you and your people don't know where you are going, your leadership doesn't matter.

—Ken Blanchard

Fire service leaders will find it difficult to grow their organization without change. Conquering such change starts with having a solid game plan.

Your fire department's personalized game plan is the road map to your organization's success. A written recruitment and retention (R&R) *game plan* shows how your organization will transform. To control your organization's destiny, you must know the direction you want to go, the resources you need to advance, and who is going to own the actions it takes to move forward. Developing this strategic plan will keep you moving forward.

Successful football teams have a written game plan all team members are familiar with, or playbook. A good playbook identifies the positions needed, defines the roles and responsibilities of each position, provides an overview of the team policy, identifies who everyone in the organization is and what they are responsible for, identifies resources, lists the time needed for training and communication of information, obstacles for each play, and the tasks (plays) needed to succeed. Fire departments need their own playbook to help them win the R&R game.

As you design your department's R&R game plan, consider the following tips:

- Your strategic R&R game plan should align with your fire department's mission.
- Design a plan that is realistic and workable.
- Include specific measurable goals (see S.M.A.R.T. goals in Chapter 6).
- Support those goals through the task actions.
- Delegate a responsible party to oversee *each* specific goal.
- Avoid setting goals that are static, with no supporting rationale.

Why Develop a Strategic Recruitment and Retention Game Plan?

While there is no one-size-fits-all approach to preparing a strategic plan, developing a custom game plan allows the fire department flexibility in choosing what outcomes are important them and how they will measure them. Before providing you with the planning tool, let's break it down to gain a better understanding of what it entails.

There are nine critical elements in the R&R game plan. Each element is designed to serve as a key part of your playbook. Moreover, each element is intended to gather key information for each specific goal you will set.

Critical Elements of the Recruitment and Retention Game Plan

The R&R game plan uses a S.M.A.R.T. goal philosophy to clarify the critical path to attainment (see chapter 6 for an introduction to S.M.A.R.T. goals). The nine critical elements fulfill each letter of the S.M.A.R.T. acronym.

The following nine critical elements should be included in your R&R game plan:

- What
- Who
- Current resources
- Needed resources
- Potential barriers
- Overcoming barriers
- Communication
- Timeline
- Progress check

You can create your own game plan on a spreadsheet. Make each element a column to be populated.

Column 1: What

The "What" column answers one simple question: What specific goal are you trying to accomplish? Down the "What" column of the R&R game plan, you should list *each* specific goal your organization hopes to achieve. These goals should not be vague, but rather should document a specific assignment needed to attain the overall goal of R&R.

One common, big mistake people make when developing their game plan is the failure to set specific goals. For instance, "Recruit more volunteers" is not a specific goal. That is your desired outcome. Your *what* should be the specific action step that you want to take to accomplish the desired outcome of recruiting more volunteers. By contrast, an example of a specific goal to help reach that outcome would be "Develop an online membership application."

Column 2: Who

This column identifies an individual, not a group, to manage and provide oversight of the specific goal. Another big mistake organizations make is spending a lot of time talking about great ideas and goals during meetings, but never assigning a *who* to lead the charge. Inevitably, by the next meeting, no action has occurred. Assigning who means assigning responsibility for getting the work done including other tasks associated with the across the row from each specific goal (each *what*).

Leaders should not be assigning themselves as the who for every action. Think back to chapter 2 and the importance of engaging stakeholders. Allowing stakeholders to take on a who role allows them to take ownership in the change process. Leaders should not just assign a who. First, talk to the individual you would like to assign the task to, or *designee*, and ensure that they are willing to take ownership. The designee should want to own the goal and have the knowledge necessary to own the responsibility.

Column 3: Current resources

In this column, identify the organization's current resources that will be used to accomplish the specific goal. As you allocate resources to your R&R game plan, examine any impacts to other areas in your organization. Ensure that a reallocation of a resource doesn't negatively impact something else.

Returning our example specific goal of developing an online membership application, the current resources identified might be either of the following:

- We currently have a paper application to use as a content guide.
- We currently have a department website to host an online application.

Column 4: Needed Resources

This column identifies any resources your organization will need to acquire to meet the specific goal. It is important that the specific goals align with current and obtainable resources. The expense of the resource and how long it will take you to obtain each resource should be included in your R&R game plan. This information may alert you to a barrier for achieving your goal or may affect your timeline. Based on the availability of your resources, some goals might have to be tabled.

Returning to our example specific goal, needed resources for developing an online membership might include the following:

- Identify software to create an online form or fillable PDF file
- Identify a process for the hosting, submission, and review of online membership applications

Column 5: Potential Barriers

In this column, list any potential barriers your organization might face. These may include people, policy, and pitfalls.

People barriers include those who will resist. It is important to identify these individuals and anticipate how they will resist change. Include both internal and external stakeholders who could be barriers to the implementation of the specific goal.

Policy barriers are rules and regulations that may derail your action. It is important to identify policy barriers that you may need to overcome through lobbying or requesting special permission. Some policy barriers may lead to a need to change a specific goal. Remember, specific goals are just supporting actions to the overall goal. It is OK to change a specific goal to ensure success.

Pitfall barriers are significant financial obstacles that can severely impact your plan. These can be a direct result of needed resources.

Column 6: Overcoming Barriers

In this column, detail your plan to overcome the potential barriers listed in the preceding column. Overcoming potential barriers takes careful planning, especially when dealing with stakeholder resistance.

To address people barriers, your plan should include the following four elements:

- Involve those affected by the change
- Offer time for stakeholders to adjust to the impacts change will bring
- Provide constant communication at an appropriate level
- Ensure leadership availability to allow voicing of questions and concerns

Column 7: Communication

This column will include the methods you plan to use to communicate with both internal and external stakeholders about the specific goal and any impacts it may have. It is crucial that information about change comes from leadership and not gossip. People dislike change because of the uncertainty between the current state and the future state. One avenue of communication is often not enough, as you need to consider how each of your stakeholders prefers to communicate.

Column 8: Timeline
In this column, assign a deadline to your specific goal. Without a deadline, it is easy for a goal to be overlooked and the action(s) to stall. A deadline for goal completion allows a tool for measurement and tracking of success for the specific goal.

Column 9: Progress Check
This final column represents your benchmark checks. Benchmarks are when the goal will be reviewed (30 days, 90 days, 6 months, etc.). Also indicate who is responsible for monitoring the progress. To ensure accountability, this should be someone other than the person listed in the "Who" column.

Organizing the Recruitment and Retention Game Plan

To organize your R&R game plan, it can also be broken down into strategy sections. Examples of strategy sections include the following:

- Onboarding
- Schedule accommodations
- Nonwage benefits
- Opportunities for success
- Health and wellness
- Feedback
- Employee recognition
- Supportive environment
- Family first
- Diversity
- Recruitment messaging
- Delivery
- Targets and follow-ups
- Job satisfaction

Under each strategy section, list your specific goal.

STRATEGY STOP

Use the Recruitment and Retention game plan template shown in figure 9–1.

The Recruitment & Retention Game Plan								
The What What goal are you trying to accomplish?	**The Who** The Team Who will take charge of it? Who will monitor?	**Current Resources** Identify current resources you have to meet the goal	**Needed Resources** Identify resources to meet the goal	**The Potential Barriers** People barriers: Who will resist? How will they resist?	**Overcoming Barriers** How will you overcome potential barriers?	**The Communication** What methods will you use to communicate to the team on this task? To the stakeholders?	**The Timeline** Goal completion date When will tasks take place?	**Progress & Success Check** How will you measure progress? What are your benchmarks?
Onboarding								
Goal1: Develop an online membership application	Who: Joe Smith; Janet Brown Monitor: Chief Anderson	We currently have a paper application we are satisfied with	Need to create an online form or fillable PDF to post on our website	Barrier 1: We currently do not know who has the skillset to create the form or PDF Barrier 2: Some members may object and be resistant to change	B1: We will survey or membership to see if a member or their loved one has the skillset to assist; We can also reach out to the local high school or college to see if a student would be interested in helping B2: Develop a plan to communicate the benefits of an electronic application (reduction in paper, reduction is submission time)	C1: The goal will be briefed to members at the business meeting C2: Team communication will take place primarily by email after an initial face-to-face meeting. C3: An email will be sent to all internal stakeholders regarding the plan C4: Upon the application going live, an email will go out to all members and information about the online application will be posted on social media	Goal: Date for online application to go live 9/1 C1: July 2 C2: Team communication should take place at least 1x a week, with the project monitor being briefed every other week C3: The email will be sent after the team face-to-face meeting C4: Week of 9/1	

Figure 9–1. Example R&R game plan template

Review Questions

1. What are the nine critical elements that should be included in your R&R game plan?

2. What are the three potential barriers to implementing your R&R game plan? (Hint: They all that start with the letter P.)

Discussion Questions

1. How can your fire department benefit from developing an R&R game plan?

2. How can you use an R&R game plan to ensure that all stakeholders have ownership in meeting the overall R&R goals?

3. Brainstorm a specific goal and complete the sections of the R&R game plan for that specific strategy section goal.

10
SETTING THE COURSE FOR ORGANIZATIONAL SUCCESS

When the frame doesn't fit, shatter the frame.

—Caroline Kelso Zook

One of the biggest mistakes fire departments make is investing in costly recruitment campaigns without a clear course in mind for their recruitment and retention journey. Organizations enter the marketing phase without having clearly defined roles and what they truly need to be successful. Typically, when I ask a department leader what they need help with, they say more firefighters. However, when I ask them to make a list of all the jobs their firefighters do to keep the organization going, the majority of them identified tasks that have nothing to with structural firefighting.

Changing the Model of Recruitment

Many fire departments have relied on their first responders to perform all the functions (emergency and administrative) within the organization. We need to change this model. We are exhausting our first responders with duties they are not interested in fulfilling. We have failed to tap into the expertise of those within our community who have a desire to serve in a non–emergency-response capacity. If we break down all the roles within our fire department and assign each as emergency response or non–emergency response, then we can recruit more members to distribute the work among.

Brainstorm Needed Roles

Hold a brainstorming session with your members to identify and document the roles that exist within your organization—both emergency and nonemergency roles. Identify what training and skills are needed to fulfill each role. Table 10–1 shows an example of a few roles identified during a brainstorming session.

Table 10–1. Sample table for role outline

Role	Area(s) of Function	Type of Role	Training and Skills Needed	Length of Commitment	Notes
Structural Firefighter 1	Fire suppression	Emergency	120 hours of fire training State certification—pass test Recertification required every 3 years with required continuing education	1-year commitment after certification, if the department pays for the training	Firefighter 1 and Firefighter 2 are both options.
Fire Inspector	Fire prevention	Nonemergency	80 hours of training before state certification State certification—pass test Recertification required every 3 years with required continuing education	1-year commitment after certification, if the department pays for the training	Inspections can be scheduled around personal obligations for those with a busy family or work schedule. Members with health issues who can no longer do structural firefighting could perform this task.
Emergency Medical Technician	Emergency medical	Emergency	296 hours of training before state certification State certification—pass test Recertification required every 3 years with required CE	1-year commitment after certification, if department pays for training	Hospital offers a night and weekend course once a year.

Position					
Paramedic	Emergency medical	Emergency	1,100 hours of training before state certification. State certification—pass test. Recertification required every 3 years with required continuing education	1-year commitment after certification, if department pays for training	Hospital offers a night and weekend course once a year.
Grant Writer	Administrative—fundraising	Nonemergency			
Social Media Manager	Administrative—public relations	Nonemergency	Knowledge of social media platforms		Local college and high school have digital marketing programs—possible intern?
Photographer	Administrative—public relations, incident support	Nonemergency but may respond to scenes. Will need PPE when on scene	Digital photography skills		State fire academy offers a free course for fire-scene photography.
Social Event Planner Coordinator	Administrative—member morale	Nonemergency			
Mechanic	Administrative—logistics	Nonemergency	Mechanic		
Fire Prevention Educator I	Administrative—public relations; fire prevention	Nonemergency	18 hours of training before state certification. State certification—pass test. Recertification required every 3 years with required CE		State fire academy offers a course for fire prevention educators.

STRATEGY STOP

Host a brainstorming session to identify fire department roles.
For each role, outline the area of function it falls under, the type of role it is, the training and skills needed to obtain and maintain the role, and the length of commitment the department requires, if any (table 10–1).

Develop Job Descriptions

Once you have identified all the roles your department needs to fill, the next step is to write job descriptions for each role. Job descriptions are the foundation of job postings and recruitment advertisements, summarizing expectations for potential employees regarding skills, training needs, and performance outcomes. Job descriptions should include a brief overview of the role being filled, how it supports the mission of the department, outline a list of key responsibilities, training requirements, qualifications, and commitment expectations. Job descriptions tell potential volunteers exactly what you need them to perform and the training necessary to fill that role. Clear written communication of expectations for a particular job will help a potential volunteer to decide if they are good fit for the opportunity based on personal knowledge or time; outlining the minimum qualifications to perform a job in the description also gives you a nondiscriminatory basis for *not* placing a particular volunteer in the role.

Writing and maintaining job descriptions is key for developing your staffing plan. Having these allows you to budget money to provide job training. Another benefit for a well-developed job description is for volunteer accountability; by contrast, a vague job description makes it more difficult for a leader to address performance issues and improvement needs. It is important to review job descriptions annually to ensure that they are up to date with the latest standards and consistent with any industry trends.

STRATEGY STOP

Develop specific job descriptions for each of the needed roles your organization identified during the brainstorming session:

- *Title*. Job titles that are specific are more effective than a broad category. Use common language and avoid acronyms or industry lingo.

- *Job summary.* The summary should include an overview of your organization and the expectations for the role. Think about the job description as a first introduction to your department and sell them on why they would love to volunteer for your organization.
- *Location.* Include the exact location where the job will take place. For example, if it is an administrative support position that can be done remotely, include that information. If it is an emergency-response position with a jurisdictional living requirement, include that as well.
- *Key responsibilities.* Listing out the core responsibilities and day-to-day activities of the role is important. Include anything unique to your department that distinguishes it from other organizations. Include position requirements to ensure that potential volunteers can determine if they are qualified to apply. Outline how the position fits within the organizational structure.
- *Qualifications.* Include the required qualifications—both hard and soft skills. Document required certifications, education, and past experience; also specify background check requirements and time commitment expected. For example, if you need emergency medical technicians (EMTs), will you be seeking already-certified EMTs, or will you provide training with the expectation that the potential volunteer completes the training within a certain time frame?
- *Benefits.* Outline any financial and nonwage benefits the role has to offer. Many volunteer organizations fail to promote the nonwage benefits their opportunity has to offer potential volunteers. There are over 1.5 million nonprofits Americans can pick from to donate their time. It is your job to make your organization stand out.
- *Time requirements.* Include any length-of-service time commitments. Because of the investment in training, some volunteer positions may require a certain length of commitment to see a return.

Prepare for Onboarding

The onboarding process is not one that fire service leaders should ignore. As previously mentioned, the highest risk of turnover occurs during the first 3 to 6 months for new volunteer firefighters.[1] New volunteers who are not satisfied with the organizational climate have a greater chance of finding a volunteer opportunity elsewhere. Organizational commitment and retention in the first few months is fostered by the quality of experiences a new volunteer experiences. Organizational commitment is tied to the level of involvement a volunteer has, how strongly the volunteer identifies with the organization, and their desire to stay engaged. The greater level of identification with a social group an individual has, the lower the chance of turnover will be.

One method to reduce newcomer turnover is to develop an onboarding process for new or potential volunteers. This process should include the dissemination of key information, a documented orientation process, a career path, and a mentoring program. When providing new or potential volunteers with onboarding information, go over each document and allow the opportunity for questions. This ensures clear expectations and can reduce turnover. Among the biggest mistakes that occur during onboarding are fire service leaders not providing all the information a member will need to know to be successful or handing the newcomer a packet of documents without explaining each document.

Onboarding Sections

Onboarding can be broken down into sections, to ensure a comprehensive process.

Introduction to the organization

Under this first section of the fire department onboarding process, the new recruit should be provided with an organizational overview, introduced to the mission, and shown the organizational chart. This establishes the cultural foundation of the fire department for the new volunteer.

Administrative procedures

Under the administrative section of the onboarding process, new recruits should complete any new-volunteer paperwork and gain an introduction to practices. New members are issued passwords, access codes or keys, and core information needed to perform. This sets the expectations for being a fire department member and builds on the aforementioned introduction to the fire department's culture.

Volunteer benefits

Under the volunteer benefits section, the new recruit is provided with all of the monetary and nonwage benefit information provided to them as a member of the fire department.

Policies and procedures

Under this section of the onboarding process, new recruits are introduced to the key policies and procedures. In addition to standard polices and operating procedures, new members should be provided with an overview of communication methods.

Training

Under the training section, new recruits are introduced to all the training requirements and shown how to access trainings. In addition, they should be made aware of future opportunities and any department procedures around training. A calendar of training should also be provided to each new recruit.

Introductions and tours

Under this section, new volunteers are introduced to their mentor(s), as well as other key personnel. New recruits are provided a tour of the facilities. They are provided time to ask question and familiarize themselves with each area and the equipment.

Benefits

Under this section, volunteers are made aware of all the benefits, both nonwage and monetary, offered to members of the fire department. These benefits should include what is offered from the local department, state association, and National Volunteer Fire Council.

STRATEGY STOP

Develop an onboarding orientation checklist.

Introduction to the Fire Department

- ☐ Fire department organizational overview
- ☐ Fire department organizational culture
- ☐ Fire department mission
- ☐ Fire department organizational chart

New Volunteer Paperwork

- ☐ Employment forms (tax)
- ☐ Background check paperwork
- ☐ Member handbook
- ☐ Firefighter physical information

Benefits

- ☐ Health, life, and disability insurance
- ☐ Retirement benefits
- ☐ Educational assistance and training

- ☐ Employee assistance program
- ☐ Stipend procedures
- ☐ Incentive programs
- ☐ Discounts

Administrative Procedures

- ☐ EMS/fire reporting software
- ☐ Computer username and password
- ☐ Department-issued email
- ☐ Department key(s) and access codes
- ☐ Department ID badge
- ☐ Purchase/reimbursement requests
- ☐ Calendar of training and special events

Key Policy Review

- ☐ Antiharassment, antidiscrimination, and antiretaliation policy
- ☐ Leave of absence
- ☐ Dress code
- ☐ Personal conduct standards
- ☐ Progressive discipline
- ☐ Security
- ☐ Confidentiality
- ☐ Safety
- ☐ Injury and exposure reporting
- ☐ Emergency procedures
- ☐ Social media policy
- ☐ Information technology policy

Introductions and Tours

- ☐ Department staff and key personnel
- ☐ Township, city, or county personnel
- ☐ Tour of facility, including:
  - » Restrooms
  - » Showers
  - » Computer room
  - » Laundry room
  - » Information and resource board
  - » Member parking
  - » Supplies

- » Dayroom
- » Kitchen
- » Gymnasium

☐ Tour of trucks:

- » Each truck should have a checklist

STRATEGY STOP

Create a clear organizational chart.
Every fire department should create an organizational chart that includes all of the positions within the organization. An organizational chart should provide a visual of the hierarchy of all positions and the progression for advancement. Providing a recruit with an organizational chart may reduce any anxiety that a new member may feel. Microsoft Word can be used to create basic organizational charts by inserting "SmartArt" (from the "Insert" ribbon) and selecting "Organization Chart" (from the "Hierarchy" menu). Canva is another tool for creating an organizational chart, particularly for one that includes photos.

Provide a Written Career Path

The fire service does not always do a good job of laying out the career path for a new recruit—from next steps up to one day becoming fire chief. Having a written career path can be a critical aid to keep members on track toward achieving personal goals. Volunteers should feel like they have a future within the organization, and it is up to fire service leaders to provide them with a clear path to achieve and progress.

A *career plan* is a sequence of positions within the same organization that enables volunteers to continue serving while attaining industry short and long-term career goals. A career plan doesn't guarantee a promotion in rank, but it provides the volunteer with a future within the organization to focus on. The skills developed as the volunteer progressing through the plan should help prepare them for future opportunities.

The following are benefits of offering a career plan:

- Creating a positive workplace culture
- Sending a message that the organization cares about the volunteer and their future

- Making volunteers feel like they are valuable and not disposable
- Motivating volunteers to work harder and progress forward

Offer an individualized career plan to each volunteer, including necessary trainings that the organization will need to provide to help toward advancement. Fire service leaders should meet individually with volunteers annually to ensure that personal goals are addressed. Conversation topics should include positions the volunteer hopes to have in the long term—not just short-term plans. Having a clear plan to follow may generate excitement among the volunteer and contribute to long-term retention.

Create Flexibility in Training and Meetings

Flexibility is key for fire service volunteers. Identify strategies to offer ways volunteers flexibility with training and duty requirements that accommodates their schedules. Being flexible will contribute to greater retention.

Accommodating the scheduling needs of your volunteers also relates to stakeholder theory by meeting their needs. Many volunteers are being forced to pick between family commitments and the fire department. This causes internal conflict and can lead to disengagement. Fire departments also limit themselves by holding training on a set day and time. This prevents those who work nontraditional shifts from having the same opportunity to volunteer.

STRATEGY STOP

Develop methods for flexibility in training and meeting requirements.

- Video record training drills. This allows firefighters who missed the training to watch the drill and schedule a time to be tested on the skill.
- Develop a training lesson plan and assign different officers to deliver the training at different times throughout the week.
- Offer hybrid meetings to allow those who cannot attend (because at home with a baby or an aging parent, out of town, etc.) to join. (Still record the meeting for volunteers who cannot attend at the set time to be able to watch it when their schedule allows.)

Listen to Internal Stakeholders for Organizational Success

You can implement every strategy in this book, but if the fire service leader fails to learn the art of listening, they will sabotage all your efforts and energy. Leadership failing to listen represents a major communication breakdown.

Develop a Volunteer-Rounding Program

Volunteer rounding is a concept I developed based on the evidence-based practice of hourly rounding in the field of nursing. Health care organizations that adopted hourly rounding as a tactic saw an increase among external stakeholders in positive perception of the organization and a decrease in unwanted behaviors.[1,2]

Implementing a volunteer-rounding program is one tactic for collecting feedback from volunteers and improving leadership perception. Volunteer rounding entails a commitment by leaders to touch base with every volunteer every week. It is more than just checking in—it means carrying out specific behaviors to achieve a set of expected results.

STRATEGY STOP

Start volunteer rounding to engage with volunteers.
It is easy for leadership to get caught up in the organizational chaos of operations and become less engaged with internal stakeholders. Volunteer rounding is a proactive, structured approach for leadership to engage with volunteers.

Steps to Establishing a Volunteer-Rounding Program

Step 1: Communicate expectations to leaders
A volunteer-rounding program will work only if there is buy-in from fire service leaders expected to participate. The first step will be communicating a plan to start a volunteer-rounding program that highlights the benefits of engaging.

Step 2: Establish a schedule
Leaders will need to make a commitment to connect with volunteers each week. Create a weekly schedule of volunteer meetings for each leader. The list of volunteers assigned to each leader should rotate periodically to ensure that leaders establish a relationship with *all* internal stakeholders over time.

Step 3: Conduct volunteer rounding
Volunteer rounding should be integrated into operations and not viewed as additional work. Reaching out to volunteers can be done on a training night, after a call, or through a phone call. To be effective, fire service leaders should practice these specific behaviors:

- Express gratitude for volunteer service
- Assess environmental satisfaction
- Perform self-performance evaluation
- Survey needs for volunteer success
- Remind volunteers of nonwage benefits
- Document concerns that surface during rounding
- Make plans to follow up if needed

Review Questions

1. What strategy can you use to identify and document the roles that exist within your organization—both emergency and nonemergency roles?

2. What information should job descriptions include?

3. Name at least one of the biggest mistakes that occurs during onboarding.

Discussion Questions

1. What roles do you need filled within your fire department?

2. What type of onboarding process are you currently using? How can it be improved?

3. How can volunteer rounding have a positive impact on retention?

11
THE POWER OF ORGANIZATIONAL BRAND AND REPUTATION

Your fire department brand's is one of the organization's most important assets; invest in it.

Organizational Brand

Branding has never been more important than in today's era of digital media. Your fire department is no longer visible only to the community it serves, now that small organizations can be globalized through technology. Fire department opportunities and news are no longer shared in small print or local news; they can now be accessed globally by search engines. Fire departments with poor brand management will struggle with retaining good volunteers and recruiting new ones.

Organizational branding comprises creating an image for and guiding the reputation of your organization. Branding also differentiates your organization from others, which is valuable as individuals make decisions on where to donate time and money.

Transparency with technology has led organizations to have tremendous opportunities. This includes when an organization is doing things right as a way to strengthen internal and external stakeholder relationships. It can also help to protect the brand of the organization should a negative incident threaten the fire department's reputation.

Creating a strong organizational brand for your fire department can lead to the following benefits:

- *Respect*. When stakeholders hear your fire department's name, they will have a positive feeling associated with it, and your reputation will grow.

- *Stakeholder advocacy.* Followers of your fire department's brand will advocate for you. Influential decision makers who support you will endorse you.
- *Collaboration building.* A pool of opportunities and partnerships will become larger and stronger for your fire department.
- *Positive culture.* A fire department that operates under a desirable brand creates a positive culture that others will enjoy being a part of and outsiders will be attracted to.

Taking Ownership of Your Organization's Brand

To bolster retention and recruitment efforts, fire departments need to use branding strategically. A fire department's brand should reflect stakeholder theory and be about service to others and department needs only. Creating a powerful brand entails taking the time and effort to be aware of your stakeholders' needs, having mastery of the environment you operate in, taking inventory of what you have to offer, and leveraging all of these to stand out.

As the world changes, your fire department's brand will also need to transform. To remain competitive with other organizations seeking donations and volunteers and stay relevant, your brand should evolve over time.

Organizational-Branding Misconceptions

There are many misconceptions when it comes to organizational branding. Some believe that by doing great work, they will automatically have a great reputation within the community. These organizations may have tremendous outcomes, but if the external stakeholders are not made aware, how will they know you do great work?

To remain competitive among the other societal organizations seeking volunteers and donations, a fire department must market the contributions it makes to the community for its value to be recognized. Many citizens of communities served by volunteer fire departments assume that these are paid positions being served by career firefighters and emergency medical technicians; they don't even know that their fire and emergency services are made up of their fellow community members volunteering their time, so this is a priority to clarify.

Another misconception is that the city or township will market the fire department's brand. A fire department must have a personal organizational brand that sets itself apart from the city or township it operates in. In other words, an organization must take ownership of their brand and develop their own network of external supporters. Your township or city has other departments they are responsible for; they cannot prioritize promoting your fire department's brand.

While one-way organizational promotion can be harmful, self-promotion that engages in two-way communication adds brand value. Fire departments need to market achievements, contribute knowledge to the community, and have a presence at the table (digitally and figuratively). Branding self-promotion is not boasting; rather, it is educating stakeholders on organizational capabilities that offer value.

Setting Brand Expectations and Values

I grew up in poverty in rural Ohio, but my adoptive dad and mom taught me to take ownership of our family brand by being competitive but treating others with equality and fairness. I was taught that you shake the janitor's hand the same way you do the chief executive officer's. I was taught to tell the truth, even if that means admitting your mistakes. These same values have carried me through life and in leadership roles.

Although we are all from the same 50-mile radius, not every firefighter born and raised in my area have the same set of values. During my two decades of volunteerism in the fire service, I have seen it all—from falsified financial records, look-the-other-way leadership, and theft of fire department funds to cheating on state testing and discrimination.

Unfortunately, moral mischief and cutting of ethical corners is not limited to the Ohio I grew up in—it exists throughout today's American fire service. A few firefighters with a dark side threaten the integrity of this industry. If the American volunteer fire service wants to survive, we must clearly establish the brand expectation and values without leaving room for ambiguity. This starts with deciding what behaviors are unacceptable.

The Cost of Ignoring Poor Behavior

The culture of the fire department is defined by the worst behavior a leader is willing to accept. Every time a fire service leader looks the other way when a firefighter is amoral, the organization is threatened. This poses a threat to reputation and can lead to volunteer disengagement.

Because of volunteer shortages, fire service leaders may fail to address poor behaviors from high-performing volunteers. Active volunteers with a history of disruptive and unethical behaviors are overlooked because of their perceived value. Their leaders are afraid to lose the person, so they allow they look the other way or say dismissively, "That is just how they are."

Failing to address poor member behavior can negatively affect the organization. Other volunteers lose confidence in the leader's ability to lead. The organization's culture experiences a negative shift as amoral behaviors that aren't curbed then become acceptable within the overall culture. Ignoring poor behavior can lead the public to develop a negative perception of the organization. Thus, internal stakeholders become unhappy and may leave, and externally, a poor public image leads to recruitment barriers.

Examples of poor member behavior include disrespecting others, discrimination, rudeness, harassment, inappropriate conversations, gossip, policy violations, and passing blame. Failing to address these disruptive behaviors will cause other volunteers to grow frustrated, function in silos, or even replicate the poor behaviors to feel accepted.

Reputation Management White Paper as Foundation

Today's fire service leaders face a variety of challenges managing budgets, personnel, and programs. Occasionally, ethical issues emerge for which there are no easy answers.

—U.S. Fire Administration[1]

Over a decade ago, the Cumberland Valley Volunteer Firefighters Association brought industry leaders together from all over the country to address the threat to the fire service's reputation posed by its own members. Collaboratively, this group of leaders developed the Fire Service Reputation Management White Paper to raise awareness of the dangers from amoral firefighter behavior. Seven behavioral issues were identified in the white paper:[2]

- Cheating on examinations designed to test firefighter knowledge
- Arson fires set by firefighters
- Theft of fire service funds
- Misuse of department-owned technology and equipment
- Inappropriate use of department facilities
- Substance abuse
- Harassment and discrimination

The members of the white paper development team proposed the development of a national Fire Service Code of Ethics as a tool to educate firefighters on the expectations. Following the expectations outlined in the white paper as a code of ethics could lead to a reduction in turnover. It is imperative that fire service leaders set clear expectations regarding what constitutes acceptable behavior.

Firefighters are public servants entrusted with the safety and well-being of their communities. A code of ethics can build and maintain public trust by emphasizing honesty, transparency, and commitment to serving the public interest. A code ensures that firefighters act in a manner that is consistent with the values and expectations of the community they serve.

The Fire Service Code of Ethics

A code of ethics is invaluable in the fire service as it promotes professionalism, guides ethical decision-making, builds public trust, ensures personal accountability, supports professional development, and fosters teamwork. It serves as a foundation for the fire service's integrity and commitment to serving and protecting communities. Thus, the Fire Service Code of Ethics can play a vital role in the volunteer fire service, establishing a set of principles and standards that guide the behavior and actions of volunteer firefighters. This document can be used to clarify expectations and as a mitigation tool for forced turnover caused by unacceptable behavior.

Integrating the Fire Service Code of Ethics promotes professionalism within the volunteer fire service. It sets expectations for volunteers to conduct themselves with integrity, respect, and accountability. This helps to maintain the public's trust and confidence in the fire service.

The Fire Service Code of Ethics can serve a volunteer's moral compass. When volunteers find themselves in challenging and high-stress situations where ethically correct decisions need to be made, the code of ethics can provide volunteer firefighters with a framework to make choices that uphold the highest standards of conduct, even in difficult circumstances.

STRATEGY STOP

Implement the Fire Service Code of Ethics.

A code of ethics encourages continuous learning and professional development among volunteer firefighters (fig. 11–1). It promotes ongoing training and education on ethical issues and dilemmas, allowing firefighters to enhance their ethical decision-making skills and stay up to date with evolving ethical standards in the profession.

It is recommended that this document be reviewed bullet for bullet with each new recruit and annually. An officer should assist with the review of this document to explain the expectations and to answer any questions. This not only ensures that behavior guidelines are set, but if a member does stray from the ethical standard establishes, the document can serve as a tool for media outreach. If a member makes a mistake that is newsworthy, the chief could state, "While I cannot comment on personnel issues or open investigations, I can say that every firefighter is required to sign and abide by the expectations set in place by the national Fire Service Code of Ethics, (and here is a copy). We do not condone any behavior that violates these industry expectations."

FIREFIGHTER CODE OF ETHICS

I understand that I have the responsibility to conduct myself in a manner that reflects proper ethical behavior and integrity. In so doing, I will help foster a continuing positive public perception of the fire service. Therefore, I pledge the following...

- Always conduct myself, on and off duty, in a manner that reflects positively on myself, my department and the fire service in general.
- Accept responsibility for my actions and for the consequences of my actions.
- Support the concept of fairness and the value of diverse thoughts and opinions.
- Avoid situations that would adversely affect the credibility or public perception of the fire service profession.
- Be truthful and honest at all times and report instances of cheating or other dishonest acts that compromise the integrity of the fire service.
- Conduct my personal affairs in a manner that does not improperly influence the performance of my duties, or bring discredit to my organization.
- Be respectful and conscious of each member's safety and welfare.
- Recognize that I serve in a position of public trust that requires stewardship in the honest and efficient use of publicly owned resources, including uniforms, facilities, vehicles and equipment and that these are protected from misuse and theft.
- Exercise professionalism, competence, respect and loyalty in the performance of my duties and use information, confidential or otherwise, gained by virtue of my position, only to benefit those I am entrusted to serve.
- Avoid financial investments, outside employment, outside business interests or activities that conflict with or are enhanced by my official position or have the potential to create the perception of impropriety.
- Never propose or accept personal rewards, special privileges, benefits, advancement, honors or gifts that may create a conflict of interest, or the appearance thereof.
- Never engage in activities involving alcohol or other substance use or abuse that can impair my mental state or the performance of my duties and compromise safety.
- Never discriminate on the basis of race, religion, color, creed, age, marital status, national origin, ancestry, gender, sexual preference, medical condition or handicap.
- Never harass, intimidate or threaten fellow members of the service or the public and stop or report the actions of other firefighters who engage in such behaviors.
- Responsibly use social networking, electronic communications, or other media technology opportunities in a manner that does not discredit, dishonor or embarrass my organization, the fire service and the public. I also understand that failure to resolve or report inappropriate use of this media equates to condoning this behavior.

Developed by the National Society of Executive Fire Officers

Figure 11–1. The Fire Service Code of Ethics developed by the National Society of Executive Fire Officers

Substance Use in the Firehouse

Substance use among volunteer firefighters is a serious and complex issue that deserves attention. The unique stressors and challenges volunteer firefighters face in their service to their community can contribute to a higher risk of substance abuse. The following factors contribute to substance abuse among volunteer firefighters:

- *High-stress work environment.* Volunteers often face high-pressure situations, deal with traumatic events, and work long hours, leading to chronic stress. In turn, this stress can contribute to substance abuse as a negative coping skill to deal with emotional and psychological challenges.
- *Secondary trauma.* Volunteer firefighters often witness traumatic incidents and are exposed to human suffering, injury, and death. These experiences can have a profound impact on their mental health and well-being, increasing the risk of developing substance abuse problems through "self-medication."
- *Peer influence.* The culture within some fire departments may normalize or tolerate substance use, making it easier for individuals to develop or maintain substance abuse habits. Peer pressure and a desire to fit in can also influence substance use.

In the United States, more than 20 million adults and adolescents have struggled with a substance use disorder in the past year.[3] Substance use disorders range from alcohol misuse to illicit drug use, including prescription drug misuse. Conditions in environments where people are born, live, volunteer, socialize, and work are *social determinants of health* (SDOH). SDOH can affect a volunteer's health, functioning, and risk engagement.

Although understudied, the SDOH in the fire service can impact the psychosomatic well-being of the volunteer firefighter. In one study, among career firefighters, 58% reported binge drinking, and 20% reported hazardous drinking behavior.[4] In another study, focused solely on women firefighters, 40% of respondents reported binge drinking, and 16.5% reported hazardous drinking behavior.[5]

Substance abuse among firefighters has been known to start as a poor coping method, while others start down the path of drug addiction after being prescribed a prescription painkiller after an injury. Because of their professional relationships with medical providers and being in position of public trust, firefighters are often prescribed opioids for pain management for job-related injuries when others might not be.

The National Safety Council has indicated that 9% of working adults have a substance use disorder, and this rate is higher in male-dominated industries.[6] While effective treatments for substance use disorders exist, there are often barriers for volunteers to access these resources. It is key for fire service leaders to be educated about substance abuse, be able to identify the symptoms, offer strategies for treatment, and develop zero tolerance of substance abuse.

Individuals are more likely to engage in risky behaviors when the organizational culture condones the behavior or when there is the perception of group support. This is described by *social identity theory*. In this theory, a group of people who categorize themselves as belonging to the same social category internalize the category's social identity and attributes to define themselves.[7]

In the fire service, identity and group cohesion are valued. However, social identity has been found to influence others toward substance use.[8] Younger fire service recruits have reported binge drinking as an attempt to fit the group prototype. Fire service leaders should model positive coping mechanisms and healthy socialization activities while prohibiting substance use in the firehouse.

STRATEGY STOP

Offer resources to help volunteers overcome substance use disorder and attain member stability.

To provide assistance for volunteers working to overcome substance abuse, fire service leaders should provide tools, education, and connections.

Addressing Substance Misuse Checklist

(Adapted from the National Safety Council "Opioids at Work Employer Toolkit"[9])

- ❐ Educate yourself and organizational leadership on the impacts of substance abuse.
- ❐ Partner with organizations specializing in substance abuse and recovery to conduct educational messaging and trainings for volunteers.
- ❐ Ask officers and trusted internal stakeholders to identify areas in which improvements or changes should be made within the organization to mitigate substance abuse among members.
- ❐ Research local community groups who address substance abuse that the fire department could support, as well as events that leadership and volunteers alike could engage in (5K races, fundraisers, etc.).
- ❐ Ensure that volunteers are consistently and frequently receiving education and communications on substance abuse, including the risks and support available to them.
- ❐ Implement a Drug-Free Workplace Policy (DFWP) and ensure that volunteers are trained on how it relates to impairment.
- ❐ Communicate with volunteers about impairment, the DFWP, and where they can go for more information.
- ❐ Provide training for officers and members to assist them with understanding substance use disorders, recognizing symptoms of impairment, and responding to an impairment-related crisis.

- ❏ Implement drug-testing procedures and establish protocols related to impairment and consequences for noncompliance with testing procedures.
- ❏ Train officers and volunteers on the required procedures and documentation when on-the-job impairment is observed.
- ❏ Openly talk about substance use disorders and emphasize that recovery is possible and likely.
- ❏ Develop a department health and wellness program that promotes work-life-volunteer balance and strategies for addressing stressful conditions that may lead to unhealthy coping mechanisms, including substance abuse.
- ❏ Prohibit alcohol in the firehouse and at events. Provide a variety of nonalcoholic options and encourage healthy after-hours activities.
- ❏ Always ensure that lines of confidential communication are open for volunteers who may be struggling with substance use disorder.
- ❏ Purchase a National Volunteer Fire Council membership ($21 currently), which provides members with five counseling sessions and a bridge to local resources.

Addressing Harassment and Discrimination

The cost of turnover stemming from harassment, discrimination, and retaliation in the workplace was $172.4 billion over the past 5 years.[10] Harassment in the fire service may lead to negative financial impacts, decreased productivity, and high turnover. Departments that are proactive in preventing and addressing harassment and discrimination have higher recruitment and retention rates.

Leadership engagement and commitment are key for developing a culture within the firehouse in which harassment is not tolerated. Fire service leaders can demonstrate this commitment by taking the following actions:

- Communicating frequently that unfair treatment is prohibited and will not be tolerated
- Providing written policies and procedures on the reporting of unfair treatment and making known the consequences
- Dedicating resources and time for volunteer training on harassment, discrimination, and retaliation
- Conducting an organizational assessment to identify harassment risk factors and taking steps to mitigate these risks
- Refraining from either participating in unethical behavior or looking the other way when these behaviors occur

- Ensuring that all officers are regularly trained on the policies and complaint process
- Addressing concerns regarding the policy, complaint system, and training
- Conducting confidential volunteer surveys to assess if unfair treatment is occurring

---

STRATEGY STOP

Actively combat harassment and discrimination.
The U.S. Equal Opportunity Commission put forth the following five strategies for mitigating harassment:[11]
- Engaged and committed leaders
- Accountability that is consistent and demonstrated
- Comprehensive harassment policies
- Accessible and trusted grievance process
- Interactive and regular training developed for the organization

---

STRATEGY STOP

Develop a comprehensive and effective harassment, discrimination, and retaliation policy that is accessible to all stakeholders and is regularly communicated.
As mentioned earlier in the book, the National Volunteer Fire Council and Women in Fire have developed the Fire Service Discrimination & Harassment Toolkit for departments to use to prevent and address harassment and discrimination (fig. 11–2). This toolkit is intended to offer a guide to the following:[12]

- Who is protected by federal employment laws
- Who might perpetrate harassment, discrimination, or retaliation in the workplace
- That conduct could be inappropriate for the workplace even if it does not meet the legal definition of unlawful discrimination, harassment, or retaliation
- What actions could constitute discrimination, harassment, or retaliation in the workplace
- What to do if you suspect that you are being targeted by harassment, discrimination, or retaliation in the workplace
- Where to find available resources

Figure 11-2. Cover of the Fire Service Discrimination & Harassment Toolkit

Theft in the Firehouse

Preventing theft in volunteer fire departments is important to maintain public trust, ensure the organization's sustainability, and protect the resources and assets intended for the community or cause being served. Scandals within organizations impact all stakeholders and lead to turnover.[13] Unfortunately, theft and embezzlement scandals in the fire service have not been uncommon news. In fact, Googling "fire chief arrested for theft" yields over seven million results. News headlines have ranged from selling property stolen from the fire department to the embezzlement of thousands of fire department dollars. Fire chiefs are not the only ones finding themselves in the news for such amoral behavior; firefighters of all ranks are included.

Theft in the fire department can leads to broken trust among the public, which can lead to future fundraising barriers. A former Ohio attorney general said, "When theft or misappropriation of assets is reported or even suggested, plummeting public confidence spreads beyond the affected organization and endangers future support for other nonprofits as well." Hence, it is crucial for fire service leaders to mitigate the risk of theft.

STRATEGY STOP

Require background, reference, and credit checks of those trusted to manage department funds.

To mitigate the risk of internal theft, fire departments must prevent those with poor decision-making skills from becoming trusted public servants. Implement a thorough screening process for *all* volunteer applicants, including criminal background checks, personal reference checks, and employment interviews. The past behaviors of an individual can inform fire service leaders of the risk for future threats.

We cannot rely on gut feeling—the "seems like a nice person" philosophy. Some of the most likable firefighters have turned out to be active criminals. We must mitigate the risk to our organization—and community—by stopping them at the door.

In particular, fire department members who are entrusted with handling department finances should undergo a credit check as part of the vetting process. Individuals with a significant personal debt and credit problems are at higher risk of unethical financial handling out of desperation.[14] Many volunteers caught embezzling have indicated that they took the money with the intention of repaying the "borrowed" money back before others noticed that it was gone.

STRATEGY STOP

Develop internal and external checks and balances of financial records and inventory lists.

One best practice is to have multiple sets of eyes on records. This allows the quick catching of both intentional and unintentional mistakes. Fire departments should also prohibit the signing of blank checks and require multiple signatures. Another good practice is to have one person responsible for the review and approval of bills and another who actually makes the payment. Ensure that the parties responsible for approving and signing are independent of one another and avoid authorizing family members and close friends from serving in these roles. Ensure that all financial transactions are supported by appropriate documentation, such as receipts, invoices, and financial reports. Establish clear policies for documenting and approving expenses and regularly review financial records.

STRATEGY STOP

Reconcile and review financial records regularly and implement internal controls.
Conduct surprise audits or spot checks to deter theft and identify any irregularities promptly. The bank statements should be reconciled monthly by someone inside and outside the organization. In today's era of digital records, a copy of records should be stored for review access and not just housed on financial sites. Each month two people who do not handle funds should review the records and sign off that they reviewed them.

STRATEGY STOP

Develop procedures for expense requests and reimbursements that must be followed by all fire department members.
Expense and reimbursement approvals should involve a review and sign-off procedure to mitigate abuse. All reimbursements should require supporting documentation to verify that the expense was approved department business. Organizational credit cards should be limited, and statements should be closely monitored to ensure that charges match approved expenses. The fire department should also have a policy in place for spending and purchase-order approvals that can be verified by the reviewer. Donations, operational checks, and other financial documents should not be sent to home addresses of members. A fire department should have an owned location, such as a post office box, that serves as receiving point of entry.

STRATEGY STOP

Conduct inventory of department property.
Inventory should be accounted for through routine inventory checks conducted by more than one person. Conducting routine inventory checks can prevent expensive equipment from disappearing and ensure that any borrowed equipment is returned.

STRATEGY STOP

Rotate and cross-train department duties.
Avoid concentrating financial responsibilities with a single individual. Separate financial tasks—such as handling cash, record-keeping, and financial reporting—among multiple volunteers to minimize the risk of collusion or fraudulent activities. Rotation of and cross-training on duties represents another way to mitigate and uncover unethical behaviors. Many organizations that have been a victim of embezzlement have reported that the person of trust had been in their position for years and was the only one in that role. In addition to having another set of eyes, cross-training ensures that a replacement is readily available for the continuity of fire department business.

STRATEGY STOP

Routinely change passwords, combinations, and locks to prevent unauthorized access.
Unfortunately, not all volunteers make it to lifetime fire department member status, and some are asked to leave. As faces change, it is important to implement a policy to routinely change passwords, combinations, and locks.

STRATEGY STOP

Watch for and report red-flag behavior(s) immediately.
The following warning signs are potential red flags:[15]

- Being territorial and secretive of financial records
- Having an unwillingness to disseminate financial information
- Implementing a change in how records are kept
- Missing financial records
- Debt lacking supportive documents
- Unexpected change in personal behavior
- Not taking vacations
- Opposing the rotating of duties
- Major variances in revenue and expenses compared to previous years

- Reluctance of an audit
- Unpaid bills and late payment fees
- Late or absent financial reports
- Extravagant purchases, trips, or other lifestyle changes

STRATEGY STOP

Encourage reporting and whistleblower protection to create an environment that encourages volunteers to report any suspicious activities or concerns about theft.
Establish and educate volunteers on confidential reporting procedures and ensure that whistleblowers are protected from retaliation.

Reputation Management for Volunteer Fire Departments Is Essential

Fire service leaders must be proactive with maintaining a positive image, attracting volunteers, and ensuring ongoing support from the community. Reputation management is an ongoing process that requires consistent effort and dedication. By prioritizing expectations, transparency, accountability, positive engagement, and effective communication, volunteer organizations can build and maintain a strong and positive reputation in the community that leads to increased recruitment and reduced turnover.

Review Questions

1. What toolkit can be used to educate employers and employees on discrimination, harassment and retaliation?

2. How can the national Fire Service Code of Ethics be used when an internal stakeholder makes a mistake?

3. How many search engine results do you get when you Google "fire chief arrested for theft"?

Discussion Questions

1. What types of checks should be implemented to prevent theft in the fire department?

2. What type of discrimination, harassment, and retaliation process are you currently using? How can it be improved?

3. What are red-flag behaviors that you should be watching for to prevent theft in the firehouse?

12

CITIZEN ENGAGEMENT FOR RECRUITMENT

Every time a family stops by to see your fire truck or ambulance, that mother or father are your potential members; so are the kids.

—Stephen Marsar, battalion chief
Fire Department of New York (FDNY)

Citizen engagement programs within the fire service offer valuable education and awareness opportunities (fig. 12–1). Participants learn about fire safety, prevention strategies, emergency preparedness, and the importance of early detection and rapid response. This knowledge equips individuals with the skills and information necessary to protect themselves, their families, and their communities. These programs can also lead to recruitment opportunities.

STRATEGY STOP

Host a citizen fire academy to promote community engagement. This will allow residents to learn about their local fire department, its operations, and the challenges firefighters face. This increased understanding in turn strengthens the connection between the fire department and its community, enhancing trust and collaboration.

Citizen fire academies can also serve as a recruitment tool. Participants may develop a strong interest and passion for firefighting and decide to pursue volunteer opportunities within the department. Additionally, the academies can contribute to the retention of current volunteer firefighters by fostering a sense of pride, camaraderie, and commitment to their profession.

Consider the following tips as you design your citizen fire academy:

- Structure your citizen fire academy over a series of consecutive weeks to maximize community participation.

- Provide clear expectations before the academy starts (attendance, dress, time frame, etc.).
- Design your curriculum based on the typical encounters your volunteers encounter. For example, if farming accidents are something the members of your department often face, schedule a mock farm accident to introduce academy members to the typical challenges posed by this type of emergency.
- Host a graduation ceremony when the academy is over and recognize the participants. Invite local media and families to attend. Provide a printed certificate signed by the fire chief and bring in a speaker to offer a few remarks. Have applications to join the department available at graduation and personally invite graduates to apply.

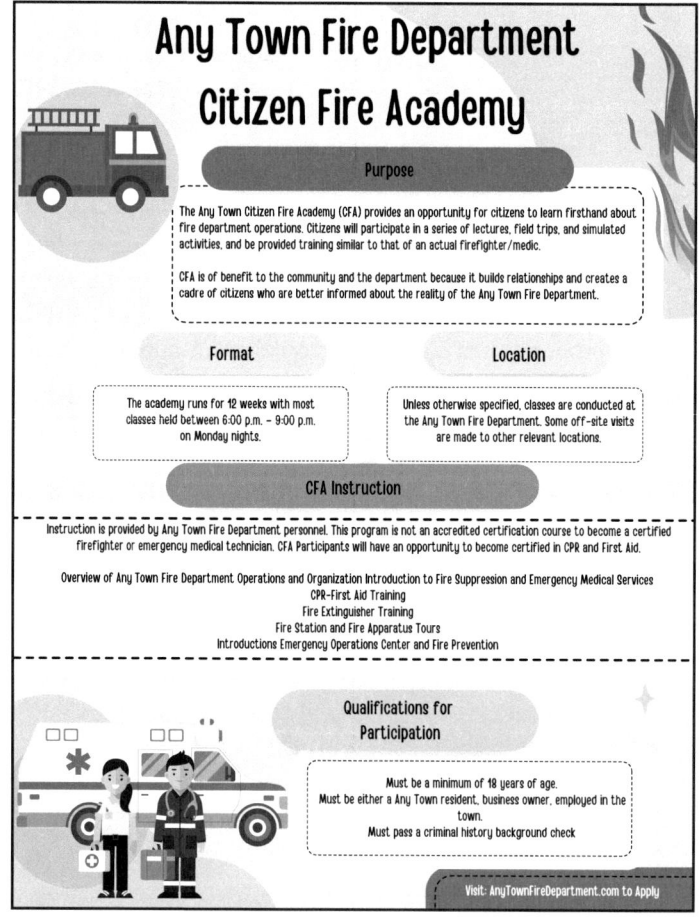

Figure 12–1. Poster encouraging participation in the Any Town Fire Department Citizen Fire Academy

STRATEGY STOP

Host "Be a Firefighter for the Day" events to allow members of the press, community leaders, and regular citizens an opportunity to come into your organization as a volunteer for the day and see for themselves the work the fire department does.

During such an event, attendees typically have the opportunity to listen to presentations, have lunch with firefighters, don fire gear, and participate in drills—search-and-rescue, auto extrication, mock emergency medical services (EMS) calls, equipment use, and so forth. These events can double as recruitment events.

Host multiple events on different dates to reach various groups. When planning a "Be a Firefighter for the Day" event, consider who you are trying to reach and educate. For example, you might host separate events for each of the following purposes:

- *Educating media.* Participating members of media would be filmed by their crews while having an opportunity to interview instructors.
- *Educating politicians.* Elected officials would be introduced to facts about the equipment they are using (cost, replacement timeline, etc.) while having an opportunity to take photographs that can be used for social media and press releases.
- *Women engagement.* Women would not only gain exposure to the firefighter profession but also a boost to their self-confidence that there is a role for them in the volunteer fire service.
- *Youth engagement.* This event would introduce youth to a future in the fire service and provide them with a clear path to get involved.
- *Veteran engagement.* This shows support to veterans—who may be looking to continue service within their local communities.

IMPACT OF FIREFIGHTER INTRODUCTION PROGRAMS

The Feel the Heat program was so valuable in providing support and confidence to women interested in firefighting. Women interested in the fire service were entirely taught by women firefighters and instructors, [which] absolutely showed them that they too can do it. My favorite part of the day came after the skills were done and you see the look of accomplishment on their faces. It was important for us to talk about the behavior, attitude and "fitting in" to the service while at the same time sticking up for yourself. What a pleasure it was to be involved in such a program during my career.

—Diane Bunker, firefighter/paramedic
EMS/fire instructor, assistant chief (ret.)
Randolph (OH) Fire Department

My start in the fire service came from two different programs introducing me to the fire service. The first one, sponsored by our county's Emergency Management Agency, allowed citizens to don fire gear and go through the search-and-rescue trailer. That initial engagement sparked my interest in the volunteer fire service, and the next program I attended was the Feel the Heat program, sponsored by the Ohio State Fire Academy. I missed my college graduation because I was so eager to attend this event. This overnight weekend program was for women and taught by an all-women staff of instructors. The weekend consisted of one day of classroom information and one day of hands-on skills in full turnout gear, to include a live burn (fig. 12–2). This program gave me the confidence to take the next step and enroll in Firefighter 1 to serve my community.

Figure 12–2. Ohio Fire Academy, Feel the Heat program, 2008 class

STRATEGY STOP

Host a youth fire camp.
This will provide an opportunity for youth to see for themselves the work that the fire department does and gain the confidence to join when they are old enough. These camps can provide valuable educational experiences, teaching young people about public safety and emergency services, while also fostering personal development and promoting community safety. That they serve as a recruitment path is an added bonus.

Chief Shelly Carter from the City of Hartford, Connecticut, Fire Department founded Girls Future Firefighter Camp (GFFC) (fig. 12–3). This camp is a training program that focuses on empowering girls 13 to 18 years old to think outside the box and open their minds to the possibility of careers in public service. GFFC attendees are provided with training in many areas of the fire department, including EMS, fire prevention and risk reduction, fire investigation, and personnel support. Participants are walked through a day in the life of a firefighter and introduced to turnout gear, PPE, and breathing apparatus. The camp is 100% free.

Figure 12–3. Poster for GFFC

The following are helpful tips for planning a youth camp:

- *Determine camp staffing needs.* Implement the appropriate supervision ratios and document your staffing plan. Draft an organizational chart to show all the required positions and lines of accountability. Provide a job description for each camp role that includes the job title, general responsibilities, specific duties, and qualifications.
- *Select camp staff carefully.* Conduct necessary background checks for all adult volunteers involved with the youth camp through the Bureau of Criminal Investigation and screen volunteers through the National Sex Offender Public Website (http://www.nsopw.gov/). Verify all volunteers' prior education, experiences, training, and character references.
- *Implement a facility and maintenance operation program.* Ensure that all camp equipment is in good and safe working condition. Restrict access to flammables, medical supplies, and other dangerous items. Provide an adequate number of toilets and have on-hand period supplies and a discreet method for disposal.
- *Create a food safety culture to prevent foodborne illness outbreaks.* Implement proper food storage and processing techniques, following guidelines from the U.S. Department of Agriculture and your local health department.
- *Establish written camp expectations and paths for communication with youth participants and parents or guardians.* This should include a discipline policy. All of the following practices are considered child abuse or neglect and should be strictly prohibited: corporal punishment, humiliating or frightening methods; punishment associated with food, rest, or isolation; and using foul or abusive language.
- *Maintain youth medical records.* Follow state requirements for camps.
- *Implement a training plan to ensure child safety and well-being.* Camp volunteers should be trained to recognize and report child abuse.
- *Develop a camp visitor policy.* Develop procedures for screening all visitors, authorizing their presence, and identifying authorized and excluding unauthorized visitors.
- *Integrate physical agility activities into the camp program.* Include firefighter functional exercises, obstacle courses, and team challenges that promote physical well-being and resilience.
- *Develop a safety plan for each hands-on activity.* Offering hands-on activities make the camp experience interactive. Organize simulated emergency scenarios, where youth participants can role-play as

responders and learn about teamwork, problem-solving, and critical thinking. Teach youth basic safety skills, such as CPR and first aid.
- *Invite guest speakers who have experienced emergencies or disasters firsthand.* Speakers could include survivors, first responders, and community leaders. Their stories would highlight the importance of public safety preparedness.
- *Incorporate community engagement activities.* Encourage youth to develop and implement a community safety project, such as a conducting fire safety inspection or creating public safety awareness campaigns. These projects will empower youth participants to apply their knowledge while making a positive impact.
- *Recognize attendees' achievements.* End the youth camp experience with a graduation ceremony or closing event. Provide certificates, badges, or other forms of recognition to acknowledge their completion of the program and encourage continued engagement with public safety initiatives.
- *Evaluate the impact of the camp and solicit feedback for future improvement.* Use feedback from participants and their families to continually improve and refine the program in subsequent years.

STRATEGY STOP

Host a fire department open house to provide community members with an introduction to the fire department.
Hosting a fire department open house is a great way to engage the community, raise awareness about fire safety, foster positive relationships between the fire department and residents, and offer a path for recruitment (fig. 12–4).

The following are helpful tips for planning a fire department open house:

- *Choose a date and time that is convenient for the community and your volunteers.* Usually this will be a weekend or evening when more people are likely to be available to attend.
- *Clarify the purpose of the open house.* It could be to educate the community about fire safety, showcase the fire department's capabilities, promote recruitment, or strengthen community partnerships.
- *Plan activities and demonstrations that will engage attendees and provide hands-on experiences.* These could include interactive fire safety demonstrations, simulated emergency scenarios, opportunities for

attendees to try on firefighting gear, obstacle courses, opportunities to use a charged hose line, etc.
- *Collaborate with and invite other local organizations.* They can contribute to activities, offer resources, or provide additional expertise.
- *Develop a promotional plan.* To spread the word about the open house, use various channels such as social media, schools, churches, local newspapers, community bulletin boards, and your organization's website.
- *Distribute handouts.* Provide informational brochures, pamphlets, or handouts about fire safety tips, emergency preparedness, and other relevant topics.
- *Schedule short fire safety presentations.* Select topics like home fire safety, escape plans, smoke alarm maintenance, and fire extinguisher use. Provide opportunities for attendees to ask questions and interact with presenters.
- *Include activities and games specifically designed for children.* Face painting, coloring stations, or a miniature firefighter obstacle course could be fun ways to engage the youngest attendees.
- *Provide light refreshments for attendees to enjoy.* Consider water, snacks, or finger foods.

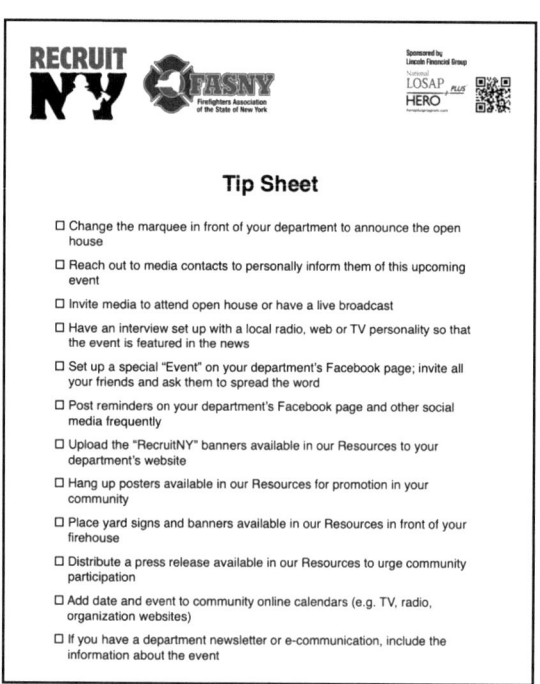

Figure 12–4. Tip sheet for fire department open houses

- *Offer branded giveaways.* These might be pens, magnets, posters, hats, or coloring sheets that reinforce fire safety messages and help participants remember the event.
- *Assign someone to document the event.* Take photos and record videos throughout. Share these visuals on social media or your department's website.
- *Make the path to recruitment visible and convenient.* Post recruitment messages throughout the event, offer applications, and hold on-the-spot interviews.

STRATEGY STOP

Host recruitment tables at existing community events to engage with citizens.

Examples of existing events include school events, festivals, fairs, church events, races, open houses for other organizations or businesses, health and wellness fairs, library events, community cleanups, park events, back-to-school events, and giveaway events (fig. 12–5).

The following are helpful tips for hosting recruitment tables:

- *Research and identify community events that align with your department's mission and target audience.* Look for events that attract a diverse

Figure 12–5. Recruitment table with information, applications, and takeaways

range of community members and offer opportunities for interaction and engagement.
- *Reach out to inquire about participation as a recruitment exhibitor.* Let the organizer(s) of the selected event(s) know that you can provide services such as on-site medical support in exchange for being allowed to participate. Discuss logistics, such as table location; whether you need to bring your own tent, table, or chairs; space and location availability for vehicles; associated fees; setup and teardown times; and any other specific requirements.
- *Create and display attractive, informative materials to showcase your fire department.* These may include brochures, flyers, volunteer position descriptions, success stories, and any other relevant materials. Ensure that the gathered materials clearly convey the benefits of volunteering with your fire department and specify how individuals can get involved.
- *Design visually appealing displays for your recruitment table.* Use gear, equipment, posters, or tablecloths that prominently display your organization's name and logo to attract traffic to your table. Incorporate infographics to highlight the impact of volunteering and the path to become an active volunteer.
- *Train volunteers who will be manning the recruitment table.* These individuals should be approachable, friendly, and knowledgeable. Encourage them to actively engage with the public by initiating conversations and answering questions.
- *Develop a brief "elevator pitch."* This will introduce your fire department and explain the benefits of volunteering succinctly.
- *Include interactive elements or activities at your recruitment table to attract attention and encourage participation.* These can spark conversations and generate interest.
- *Have a sign-up sheet or electronic device available for interested individuals to submit contact information for follow-up.* Have a laptop computer or tablet available for individuals to apply immediately and offer printed volunteer application forms that they can fill out and submit on the spot.
- *Offer small giveaways or incentives.* These serve as a token of appreciation for individuals who express interest in volunteering, as well as a reminder of your organization's contribution to the community.
- *Assign someone to document the event.* Take photos and record videos throughout. Share these visuals on social media or your department's website.

- *After the event, follow up with individuals who expressed interest in volunteering.* Reach out via email or phone call to provide more information, answer any questions, and guide them through the next steps in the volunteer application process.
- *Assess the effectiveness of your recruitment efforts.* Track the number of inquiries, sign-ups, or completed applications resulting from the event. Use this data and other general feedback to refine your approach for future recruitment tables and improve your overall volunteer recruitment strategies.

STRATEGY STOP

Offer to support local sporting events with EMS and have an informational and recruitment table at the event.

Urban and rural communities are full of school and recreational sporting events. Examples include sports, 5K races, charity walks, and trainings camps. Volunteer fire departments with EMS can ensure the safety and well-being of the participants and spectators by providing on-site EMS. Volunteering to provide EMS at events allows volunteer departments visibility and access to participant events.

The following are helpful tips for supporting local sporting events:

- *Inform the event coordinator that the fire department can provide necessary equipment and supplies for emergency medical needs.*
- *Confirm start and end times.* Ensure that you have EMS volunteers to staff the event for the full duration.
- *Obtain ideal placement.* Ask the event coordinator to be placed in a location where fire department volunteers can easily interact with the public.
- *Bring a tent and table.* Set out safety information and applications.
- *Be sure to conduct volunteer trainings and briefings prior to the event.* Ensure that volunteers are familiar with the event layout, emergency protocols, and any specific risks associated with the sport.
- *Establish communication procedures and clarify roles and responsibilities.* Review these with attending volunteers before the event.
- *Assign someone to document the event.* Take photos and record videos throughout. Share these visuals on social media or your department's website.

STRATEGY STOP

Leverage Community Emergency Response Teams as a pool of potential recruits.

Many communities offer different programs for citizens to volunteer with to prepare for and provide a basic response to a disaster within their community. These citizens are not trained first responders, but rather individuals trained with the knowledge to assist their families and neighbors until additional help is available. The goal of these programs is to empower citizens to sustain themselves independently, to allow first responders to focus on critical infrastructure and life safety emergencies. The training for these programs is minimal and can also serve as a path to recruitment for the fire service.

The *Community Emergency Response Team* (CERT) is a program under the Federal Emergency Management Agency (FEMA), introduced nationally in 1993. CERT's goal is to educate community volunteers about hazards and disaster that may affect their communities. The program has been integrated in all 50 states, with more than 600,000 volunteers being trained and over 2,700 local programs in existence. CERT supports the National Preparedness Goal: "A secure and resilient Nation with the capabilities required across the whole community to prevent, protect against, mitigate, respond to, and recover from the threats and hazards that pose the greatest risk."[1]

The CERT program offers a consistent approach to volunteer training and hands-on practice. This includes the following areas:

- Fire safety
- Response to human-made, technological, and natural hazards
- Basic search and rescue
- Team organization during basic disaster response
- Disaster medical operations

FEMA conducts or sponsors CERT Train-the-Trainer and Program Managers courses for fire, medical, and emergency management entities. Local government entities serve as a sponsor for CERT, such as a fire department, law enforcement organization, or emergency management agency. The sponsoring entity is responsible for volunteer management, oversight, training, and deployment. CERT programs are structured so that the sponsoring organization has the flexibility to create and utilize their program to meet specific community needs.

While CERT volunteers are provided with structured training for responding during an emergency, they are taught to deploy only when directed by their sponsoring entity and in a manner that complies with local procedures for their area. However, the skills gained from the training allows them to act as a good Samaritan at other times by using their skills in specific situations that occur in their own lives. Some communities use CERT volunteers for the following purposes:

- Run shelters
- Assist with basic first aid
- Shut off utilities
- Install smoke alarms
- Disseminate safety information
- Large-scale community exercises
- Staff special events within their community

Ask existing CERT coordinators if you can serve as a guest speaker at CERT meetings or invite CERT volunteers to participate in joint training. During these interactions, ensure that CERT members know there is an opportunity for them to also volunteer with your department.

Postinteraction Follow-Up

Developing a plan to follow up with interested citizens is critical to maintain their engagement and convert them to actual volunteers. The following are strategies to navigate the follow-up process:

- *Have a rapid response.* Respond to the applicant or potential applicant as soon as possible after receiving their application or verbal expression of interest. This response can be an email, text, or call thanking them for their interest and explaining the next steps. There should be a call to action to connect with them again.
- *Customize follow-up communication.* Personalize the experience by addressing each individual by their preferred name, and reference details from the previous interaction or their application. This demonstrates that you value their individuality and are eager to connect with them.
- *Provide additional information during your follow-up communication.* Share information about your fire department, its mission, and the specific volunteer roles or opportunities available. Inform them of the time frame and training required to become an active volunteer. This gives the individual a better understanding of what they can expect and provides them with the ability to make an informed decision about volunteering.

- *Outline the next steps in the onboarding process.* Let individuals know what they can expect after submitting their application. Inform them of necessary background checks, interviews, training sessions, or orientation programs that they will need to complete. This sets expectations and helps to reduce frustrations.
- *Encourage individuals to ask any questions or address any concerns they may have.* Provide contact information or a point of contact who they can reach out to for further clarification.
- *Establish a system for regular follow-ups with the applicants as they are going through the formal process.* Send periodic emails or texts or make phone calls. This shows that you are invested in their volunteer journey and value their commitment.
- *Extend personal invitations to events, trainings, orientations, or other activities related to your department.* Inviting applicants to participate in these activities allows them to experience your organization's work firsthand and connect with current volunteers.
- *Express gratitude for the individual's interest.* Recognize their commitment and emphasize the positive impact they can make within the community by becoming a volunteer. Genuine appreciation builds a positive relationship and fosters a sense of belonging.
- *Maintain organized records of your follow-up communication* (fig. 12–6). This allows you to track their progress, note any specific requests or considerations, and ensure that no one falls through the cracks during the onboarding process.

Applicant	Address	Phone	Email	Area(s) of Interest	Assigned Mentor	Date of Follow-Up	Outcome/ Notes
Jenny Recruit	1234 Fake St., AnyTown, USA	867-5309	Jenny@ IGotYourNumber. com	Fire EMS Education	Tommy Tutone	11/2/2022	Scheduled to meet at Station 1 on 11/7/2022 for a tour

Figure 12–6. Sample applicant tracking spreadsheet

Review Questions

1. What type of checks should be conducted of volunteers staffing camps?
2. How should you document events your department hosts or participate in?
3. What organizations should you consult for food safety guidance?

Discussion Questions

1. What sporting events take place in your community where you might have a table for EMS and recruitment?
2. At what events that take place in your community could have a table?
3. What should you include in your department's internal training for staffing community events?

13

PREPARING FOR OUTREACH

Every member of your fire department is a recruiter.

—Stephen Marsar, battalion chief
Fire Department of New York (FDNY)

Fire service leaders need to be prepared for outreach. Effective volunteer outreach requires consistent effort, creativity, and ongoing evaluation. Outreach will raise awareness about your department's mission and establish meaningful connections with community stakeholders.

An *outreach plan* is essential for volunteer recruitment. This establishes a structured and strategic approach to engage with potential volunteers. A well-developed outreach plan increases the chances of attracting committed volunteers who can contribute to the success and sustainability of your department's programs and initiatives.

Goals versus Objectives

Good outreach plans have goals and objectives. Developing goals and objectives effectively will help your outreach plan be successful. Goals and objectives are not the same thing, and the difference between them can be confusing. Examples of both are given in table 13–1.

A *goal* is a desired state that an organization envisions, plans, and commits to obtain. Goals are broad and provide a focus for planning. Goals are the warm and fuzzy cheerleading phrases that propel the team forward.

Objectives are the rungs in the ladder toward goal achievement. Each rung is a realistic target for the plan. Objectives should be drafted in an active tense with strong verbs such as "plan," "write," "conduct," and "produce." Objectives answer the questions of the who, what, when, and why and to what standard.

Table 13–1. Example goal and objectives

Goal example	The goal of this project is for Any Town Fire Department is to quickly onboard new members.
Objective examples	By November 1, 2024, the Any Town Fire Department R&R Committee will observe, analyze, and report the average amount of time it takes to onboard a new recruit.
	By November 15, 2023, the R&R Committee will survey new recruits to determine the acceptable time for moving to a member and suggested activities to make the transition time more acceptable.
	By January 1, 2024, the Any Town Fire Department will vote to select three steps that they will implement to reduce the conversion time for applicants.
	By June 1, 2024, there will be a 25% decrease in the average applicant wait time.

Start by defining your outreach objectives. What are the objectives of your volunteer outreach campaign? What specific outreach objectives do you want to achieve? Do you need volunteers for specific roles? Understanding your objectives will help you to design an effective outreach plan.

Determine the target audience for your volunteer outreach efforts. Consider the demographics, interests, and motivations of potential volunteers, which is also known as *target segmentation*. Are you targeting specific age groups, professionals, students, or community members? Tailoring your messaging to the intended audience will increase the chances of attracting interested volunteers. By understanding your target audience and connecting your messaging to their interests and motivations, you can effectively engage with the right people and increase the chances of attracting committed volunteers.

An outreach plan helps you to craft compelling narratives and stories about the impact of volunteering with your fire department. Share success stories and testimonials from volunteers. Stories of volunteers making a difference can inspire potential volunteers and make an emotional connection with them. Engaging storytelling creates a sense of purpose and encourages individuals to get involved.

STRATEGY STOP

Select an effective outreach tool to reach with the right audience for the message you wish to share.

Effective outreach starts with selecting the right tool (fig. 13–1). To select the right tool, ask the following questions:

- What is the outreach goal?
- Who is the target audience you want to reach?

- How can you reach your target audience?
- What barriers might you face?
- Do you have the resources to carry out the selected tool (money, people, technology, talent, etc.)?

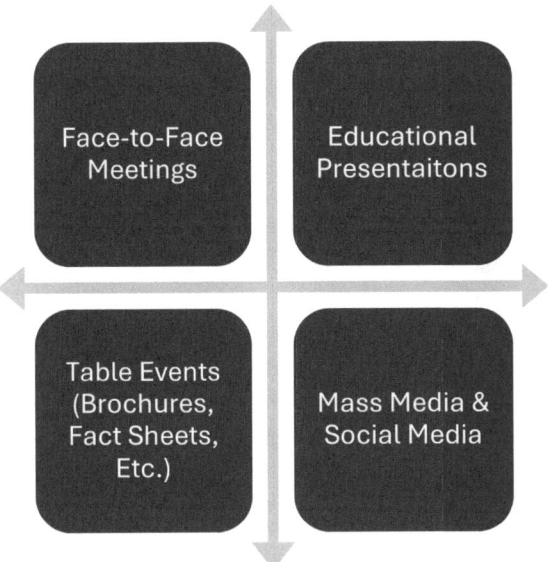

Figure 13–1. Different types of outreach

Promote the Impact of Volunteering through Positive Messaging

Craft compelling and positive messages. Clear and persuasive messaging will highlight the benefits and impact of volunteering with your fire department. Emphasize the value that volunteers bring to the community, the personal growth opportunities, and the positive change they can contribute to. Highlight the personal impact that volunteering with your organization has had on your members.

Avoid negative messaging, such as "Volunteer or else our doors will close!" As recruitment subject matter expert Tiger Schmittendorf says, "No one wants to join the Titanic Fire Department." Why would stakeholders—or anyone—want to support a sinking ship? Often departments send messages to the community that they are desperate for help, can't find volunteers, and are in danger of closing their doors (fig. 13–2). They paint a doom and gloom picture—a "Titanic" volunteer opportunity—instead of selling potential volunteers and

Figure 13–2. Flyer advising on what to avoid when recruiting volunteers

stakeholders on a beneficial opportunity for someone who makes an investment (in time or money).

Outreach Message Developing

Targeting is key for effective outreach. In other words, messaging should have an audience in mind. A targeted approach ensures that your department invests outreach resources in areas that may lead to qualified recruits.

STRATEGY STOP

Create impact stories as outreach tools to show the positive impact of volunteering.

Ask your members how joining the fire service has impacted their life in a positive way and promote their story as a department success story. Your members are representatives of the community being served. Be sure to conclude every impact story you share with a call to action.

Sample Impact Story

The volunteer fire department not only provided me with an opportunity to serve my community, but it also provided me with a path to my career. The skills Any Town Fire Department provided me led to a full-time job as a patient care assistant at Any Town Emergency Department. If you are looking for a job in healthcare, I encourage you to volunteer with us at Any Town Fire Department to gain job skills for your future.

—Jane Volunteer, Firefighter/EMT

To join Any Town Fire Department, click here. [This is your call to action.]

STRATEGY STOP

Capture the value volunteers contribute internally and highlight those contributions during fire department meetings and in internal written communications.
The following are examples of ideas to showcase:

- Volunteers who donated an item or service outside of their typical volunteer roles, such as food, a tool, product, or service (maintenance or repairs)
- Volunteers who attended local government meetings or wrote letters to elected officials on behalf of your organization
- Volunteers who established partnerships with other service organizations
- A positive public relations story that mentions the subject as being a volunteer member of your fire department, even if the article was not focused on your department itself
- Volunteers who taught a fellow member how to use new software, a new tool, or a general life hack

Most volunteers do not like to brag about the things they do or have done for the organization. Still, we need to encourage our members to share the internal value they offer. Survey your volunteers in person and through online communications frequently to collect this information.

Sample Volunteer Survey Questions

- Did anything happen this month while volunteering that made a positive impact on you?
- Did you donate any services or items this month beyond emergency response officer requirements?
- Did you attend any external meetings or trainings this month?

- Did you establish any new community contacts?
- What good things happened in your life this month that you would like to share?
- Did you teach a fellow member a new tool?
- Did you recruit any new volunteers?
- Did you participate in any outreach events for the fire department?

Develop a Plan for Postresponse Engagement

Following up with individuals following an emergency response to their location is one outreach method for external stakeholder engagement. Fire departments should create an official system for reaching out postresponse (table 13–2). This tactic should become part of your organization's culture and an extension of service. Setting goals for postresponse engagement allows fire departments to accomplish several positive aims:

- Increase compliance with any safety information provided
- Improve life safety outcomes
- Reduce frequent responses to the same address
- Decrease complaints
- Increase stakeholder satisfaction

Table 13-2. Examples of postresponse engagement

Behavior	Sample Phrases
Empathy and concern	Mr. Jones? Hello, this is Firefighter Jane from Any Town Volunteer Fire Department. I understand some of my coworkers were at your house yesterday. I just wanted to call to see how you were doing today and if you had any questions about the service we performed.
Life safety outcomes	Do you have any questions about the small fire we put out in your kitchen? Do you have and know how to use a fire extinguisher in the event of another cooking incident? Do you have working smoke alarms throughout your house? We can bring you some and install them if you don't. Do you check monthly to make sure they work?
Reward and recognition	We like to recognize our volunteers who provided excellent service. Is there anyone who you felt did an exceptional job while at your home? Can you me what they did that made you feel it was outstanding service?
Service	We want to make sure you were satisfied with our service. Were you satisfied?
Process improvement	Do you have any suggestions for what our volunteers could do better the next time they respond to a similar incident?
Appreciation	Thank you for taking the time to talk with me today. Do you have any questions?
Call to action	Before we hang up, I wanted to let you know that Any Town Fire Department is actively recruiting volunteers to serve in emergency and nonemergency roles. We need volunteers to help serve as firefighters and medical first responders. We provide all of the training at no cost to the volunteer. We are also in need of administrative support, such as volunteers to help with our website, fundraisers, and our outreach program. Would you be interested in learning more about ways to get involved with us?

Review Questions

1. Good outreach plans have _____ and _____.

2. Objectives answer these types of questions.

3. Effective outreach starts with _____.

Discussion Question

1. Present and discuss your sample personal impact story.

14

INCORPORATING SOCIAL MEDIA

Don't talk about it, be about it.

—Stephen Marsar, battalion chief
Fire Department of New York (FDNY)

Social media quickly became the most effecting marketing tool of the millennium. In 2021 there were over 4.26 billion users of social media around the globe, with a projected increase by 2027 to almost 6 billion.[1] Social media platforms provide the largest audience for engagement and a direct connection to external stakeholders.

It used to be that marketing efforts were controlled by marketing professionals and media; however, with social media, marketing now belongs to everyone. Some fire service leaders have resisted fully embracing social media and fail to see the benefits. By contrast, though, outside of the fire service, many other service organizations have learned to yield the vast power of social media to increase stakeholder support and bolster recruitment and retention efforts.

The fire service is competing with an endless list of organizations offering a platform for civic engagement. As a volunteer opportunity for community members, the fire service must do more than just show up—they need to stand out. Social media can help fire departments to not only gain supporters (followers) but also establish meaningful relationships (engagement).

Cultivating engagement is a process. Followers need to progress up the engagement ladder for fire departments to see the benefits. There are immense and rewarding benefits to using social media when the proper strategy is in place. The following are key benefits of social media use in the fire service:

- Curating a community of supporters
- Ability to build relationships with minimal resources

- Fostering a sense of belonging among external stakeholders that can be transformed into action (fundraising, volunteering, advocacy)
- Generating awareness about fire department actions
- Creating advance fundraising and corporate sponsorship activity
- Attracting stakeholders through storytelling content
- Extending your department's reach beyond a mailing list (print or digital)
- Promoting virtual and in-person fire department–sponsored events

---

STRATEGY STOP

Determine the social media platforms that your fire department will engage in to meet specific demographics.

Choose social media platforms based on the message intent and demographics you are trying to reach (table 14–1).

Table 14–1. Social media platform statistics[2]

Platform	Monthly Users Worldwide	Largest Age Group (% of Users)	Sex	Average Time Spent by User, in Minutes
Facebook	2.96 billion	25–34-year-olds (29.9)[a]	44.0% female, 56.0% male	30.0
Instagram	2 billion	18–24-year-olds (30.8)	48.2% female, 51.8% male	30.1
TikTok	834.3 million	18–24-year-olds (21.0)	54.0% female, 46.0% male	45.8
X (formerly Twitter)	237.8 million	18–29-year-olds (42.0)	34.1% female, 61.3% male	34.8
LinkedIn	930 million	30–39-year-olds (31.0)	43.0% female, 57.0% male	NA[b]
Snapchat	750 million	18–24-year-olds (34.0)	51.0% female, 48.0% male	30.4
YouTube	2.1 billion	15–35-year-olds (highest reach)	51.4% female, 48.6% male	45.6

a. Of people over 65 years of age, 56% are active users.[3]
b. NA = not available; however, 22% of users access daily.

STRATEGY STOP

Implement a content calendar to plan your social media.
A *content calendar* is a collection of your upcoming social media posts organized by date and time. This calendar can be a spreadsheet, digital calendar, or interactive dashboard. A social media content calendar serves as a cheat sheet, ensuring that your department is planning quality content on a regular basis and lessening the possibility of missed post dates. The calendar also allows departments to dates and topics not affiliated with the department. For example, if National Pancake Day is coming up, you can plan a post using pictures of a past fundraiser pancake breakfast and sharing dates of the next event.

Your social media content calendar should include the following:

- Date and time your post will go live
- Which social media account each post will be used for
- Supportive materials, such as copy, hashtags, graphics, and links

Create a mix of social content by using a variety of content (videos, text-based posts, infographics, live streams, testimonials, etc.). In particular, some platforms are ideal for certain content:

- Instagram for visual storytelling
- YouTube for how-to videos, such as "How to Conduct a Home Safety Check"
- LinkedIn for posts that engage the community with conversation
- X (formerly Twitter) for real-time updates, such as road closures

Your department can use the content calendar to do the following:

- *Plan posts in advance.* This gives time for any necessary leadership approval.
- *Ensure that posts are error free.* Have a second set of eyes review the planned post.
- *Offer timely content.* This allows departments to create content tied to current events (e.g., Christmas Tree safety during December).
- *Note key events.* One example would be fire prevention week.
- *Keep relevant with the latest hashtags.* Monitor posts with pertinent hashtags and find opportunities to use them.
- *Collect visual material.* This enhances and draws attention to the post.
- *Track social media metrics.* Use these to recognize engagement, understand trends, and plan future posts.

STRATEGY STOP

Use artificial intelligence to develop content for all your social media.
Artificial intelligence (AI) has changed the landscape of social media for volunteer organizations. Fire service leaders can use AI technologies to develop captions and write social media posts.

OpenAI's ChatGPT is free chatbot technology (available to sign up at https://openai.com/chatgpt) that can generate humanlike, conversational responses. Fire departments can use this technology for education, recruitment, and community engagement.

Importantly, AI-generated posts must be reviewed for appropriateness and edited for accuracy, before being shared with the community. In addition, when creating social media posts for recruitment, remember to include eye-catching visuals, such as photos or videos of firefighters in action, and use relevant hashtags to increase visibility. Good luck with your recruitment campaign!

AI-Generated Social Media Posts

Let's look at some sample social media posts created with the help of AI.

Text entered in the chatbot dialogue: "witty fire department recruitment social media posts"

Chatbot responses:

- Ready to turn your passion into a career? Join the hottest team in town and become a hero!
- Fire Department Recruitment is now open! Apply today and let's spark your future together. #JoinTheFireSquad #IgniteYourPotential
- Want a job that's always on fire? Become a firefighter and face the heat head-on! We're looking for brave souls to join our elite team. Apply now and let's make a difference together! #FuelYourPassion #FirefightingHeroes
- Rise to the occasion, extinguish the doubt, and ignite your potential! Join our dynamic Fire Department and make a real impact on the community. Apply today and let's set the world ablaze with our passion for service! #FuelTheFlame #JoinTheFDFamily
- Sound the alarms, it's recruitment time! We're searching for dedicated individuals who have the courage to face any challenge. Think you've got what it takes? Apply now and prove your mettle! #ReadyToServe #FirefighterRecruitment

- Calling all heroes in the making! The Fire Department needs YOU to join our ranks and be part of something extraordinary. Step up and answer the call today! #IgniteTheHeroWithin #JoinTheFireSquad
- Embrace the challenge, embrace the thrill! Firefighting is more than a job; it's a way of life. If you're ready to make a difference and save lives, apply now and let's blaze a trail of excellence together! #JoinTheBravehearts #FirefighterLife
- Ready to put out fires and light up hearts? Join our Fire Department and become a beacon of hope for the community. We're seeking passionate individuals who are eager to make a positive impact. Apply today and let's shine bright together! #IgniteTheCommunity #FirefighterJobs
- Fire Department Recruitment Alert! If you're looking for a career that's action-packed, challenging, and rewarding, look no further. Join our elite team of firefighters and let's create sparks that change lives! #FuelYourAmbition #JoinTheFireForce
- Heat up your future with a career that matters! Become a firefighter and experience the thrill of saving lives. Apply now and be part of our courageous Fire Department family! #IgniteYourFuture #FirefighterRecruitment
- Are you fueled by courage and a desire to serve? The Fire Department wants YOU! Join our incredible team and let's work together to protect and serve our community. Apply today and let your bravery shine! #FuelTheCourage #JoinTheFireHeroes

Text entered in the chatbot dialogue: "social media post from the fire department on fire safety"

Chatbot response:

> SAFETY FIRST!
>
> Hey there, our amazing community! Your friendly neighborhood Fire Department here, dropping some FIRE SAFETY knowledge to keep you and your loved ones protected!
>
> 1. CREATE AN ESCAPE PLAN: Prepare for the unexpected! Make sure everyone in your household knows the quickest way out of your home in case of a fire. Practice it regularly, because practice makes perfect!
> 2. CHECK YOUR SMOKE DETECTORS: These little heroes save lives! Test your smoke detectors every month, and don't forget to change the batteries at least once a year. Trust us, they're worth it!

3. WATCH YOUR COOKING: The kitchen is where the magic happens, but it's also a hotspot for potential fires. Never leave cooking unattended and keep flammable objects away from the stove.
4. SAY NO TO OVERLOADING: Avoid plugging too many appliances into one outlet or power strip. Overloading can lead to overheating and, you guessed it, fire hazards!
5. SMOKING SAFETY: If you smoke, do it outside. Always extinguish cigarette butts properly, and NEVER smoke in bed. Let's keep our bedrooms cool without the need for actual flames!
6. CANDLES WITH CAUTION: Candles are cozy, but they can also be dangerous. Always keep them away from curtains, paper, or anything flammable. Better yet, opt for flameless candles!
7. GRILL SMART: BBQ season is here! Keep your grill a safe distance from your house and any flammable materials. Remember to shut off the gas when not in use.
8. TEACH KIDS ABOUT FIRE SAFETY: Knowledge is power! Educate your little ones about fire hazards and what to do in an emergency. It's never too early to learn life-saving skills!

Safety is a TEAM effort! Share this post with your family, friends, and neighbors to spread the awareness. Let's work together to prevent fires and keep our community safe!

Stay safe, everyone! If you ever need us, we're just a phone call away. #FireSafety #CommunityFirst #StaySafe

HootSuite Caption Generator
HootSuite is a free caption generator tool (https://www.hootsuite.com/social-media-tools/ai-caption-generator-social-media) that can be customized based on the selected social media platform. It is easy to use, with five quick steps:

1. Indicate network (platform; e.g., Instagram, Facebook, Twitter, LinkedIn, TikTok, Pinterest)
2. Select style (choose among more than 30 options, from casual to used-car salesperson)
3. Choose language
4. Provide a brief description
5. Enter up to four keywords

Network: Instagram
Style: Educational
Language: English

Description: Picture of firefighters installing a smoke alarm
Keywords: Fire safety, fire prevention, smoke alarms, home safety
HootSuite captions:

- Did you know that smoke alarms can save lives? Watch these firefighters in action as they install a smoke alarm, ensuring the safety of this household. #FireSafety #SmokeAlarms #HomeSafety
- Learn about the importance of fire prevention with this informative image! Firefighters are hard at work ensuring every home is equipped with a smoke alarm, a crucial step towards ensuring your family's safety. #FirePrevention #Educational #HomeSafety
- Fire safety starts at home! Witness the dedication of these firefighters as they install a smoke alarm, a simple yet effective measure to protect your loved ones. Stay informed and protected with proper fire prevention measures. #FireSafetyTips #HomeSafety #KnowledgeIsPower

Network: Instagram
Style: Playful
Language: English
Description: Picture of firefighters installing a smoke alarm
Keywords: Fire safety, fire prevention, smoke alarms, home safety
HootSuite captions:

- These fearless firefighters are literally bringing the heat to fire safety! Check out how they expertly install a smoke alarm for maximum home protection. #FirePrevention #SmokeAlarmInstallation #SuperheroesInUniform
- Home safety just got a whole lot hotter! Watch as these brave firefighters save the day by installing a smoke alarm. Don't wait for danger to strike, be prepared! #FireSafetyFirst #MakeYourHomeFireproof #FirefightersAreOurHeroes
- When it comes to fire safety, these courageous firefighters are on top of their game! Check out this awesome shot of them installing a smoke alarm, truly putting their expertise to work. Stay safe and secure at home! #FirePreventionMatters #SmokeAlarmHeroes #StayAlertStaySafe

Network: Instagram
Style: True crime podcast host
Language: English
Description: Picture of firefighters installing a smoke alarm
Keywords: Fire safety, fire prevention, smoke alarms, home safety

HootSuite captions:

- Attention, true crime enthusiasts! Did you know that smoke alarms are the unsung heroes in preventing tragic fires? Learn the secrets of fire safety and home protection on our latest episode. Tune in now! #FirePrevention101 #HomeSafetyHacks
- Hold onto your detective hats, folks! We're diving deep into the world of fire safety and unraveling the importance of smoke alarms. Don't miss our jaw-dropping revelations and expert tips to keep your home secure. Detect and protect! #UnmaskingFireSafety #CrackingTheSmokeAlarmCode
- Calling all armchair detectives! Our latest podcast episode is a thrilling exploration of fire safety and the vital role of smoke alarms. Join us as we uncover the hidden truths and share powerful insights on protecting your loved ones. Get ready for an eye-opening journey! #UnlockingFireSecrets #SaveLivesWithSmokeAlarms

---

STRATEGY STOP

Navigate online criticism on your social media.
All organizations—government, nonprofit, and for-profit—face criticism at some point. This can come from internal and external stakeholders. Feedback with the right intent can be valuable and lead to necessary change. How you respond to public feedback can have a direct impact on your organization.

The following tips can help you to successfully navigate online feedback:

- *Avoid a defensive mindset.* Respond to feedback—don't react to it. For example, if someone posts that they had a negative experience with one of your firefighters, respond with empathy and a path for listening: "We are sorry to hear you feel that way. The Any Town Fire Department takes pride in serving our community. We would like to invite you meet with us so we can learn more about your situation."
- *Take the conversation offline.* Solving the problem publicly is not always productive. Take the conversation offline, but first let all readers know you care. "We are sorry to hear about your experience. We have sent you a direct message to learn more about the events you experienced, so that we can properly explore this issue. Please let us know if you didn't receive our message."
- *Own your mistakes.* Every organization makes mistakes: because all people are fallible, human errors can happen. When these inevitable

mistakes are in the public eye, offer a plan for corrective action. For example, "We are aware that we were unable to provide a crew for the call that occurred at the middle school on Monday, and the mutual aid department had a hard time locating the exact call. Many of our volunteers were still helping another family from an earlier call, which is why we relied on mutual aid. To correct this in the future and prevent delays, we have ensured that the neighboring department is aware of what drive and door to enter. We also have openings for new volunteers to serve the community, during the day. If you are interested in learning more, visit AnyTownFireDept.com."

Develop Social Media Policies

Develop a social media policy that outlines rules and guidelines for members to follow when posting content. In particular, note any required branding information.

The security of social media accounts should be a priority for fire service leaders. Communicate the importance of keeping passwords and accounts secure. Guidelines should be implemented for logging in and out of organizational accounts and where passwords are stored.

Identify who the responsible individuals will be for each social media account. Also identify who will be the administrator managing all the account users. Importantly, account administrators should have a department-issued email connected with the social media accounts. This ensures that if something happens to the administrator, the organization can continue to access the social media accounts. Ensure that there is a clear chain of command regarding social media accounts and rules to protect the organization's reputation.

STRATEGY STOP

Develop a social media policy for your organization.
Actions to consider when developing your social media policy include the following:

- *Implement a social media committee.* This group will be responsible for researching social media policies, drafting the organization's policy, getting it approved by legal, and presenting it to the membership.
- *Set the tone for the organization.* Set the expectation for the tone and voice of all posts to ensure the posts are tied to the organization's

mission, vision, and values. Restrict users from engaging in controversial topics, such as politics, religion, and other highly debated issues.
- *Create a system of checks and balances.* Clarify who has access to what platform and where the login information is stored. Create a system to review posts before they go live—more than one set of eyes should always review any post before it is visible to the public.
- *Decide how the policy will be enforced.* This includes who will have that responsibility to investigate and enforce infractions.
- *Obtain consent.* Share information about volunteers, community members, or donors only after explicit consent has been given.
- *Follow copyright laws.* Ensure that you have explicit permission when sharing content, whether textual or visual. Give credit for photos and videos shared from those outside of your organization.

Review Questions

1. How many social media users were there worldwide in 2021?

2. What is a social media content calendar?

3. OpenAI's ChatGPT is a free chatbot technology that has the ability to generate what?

Discussion Questions

1. What procedure(s) can your department develop for posting social media content of training, calls, events, and member news?

2. What events in your community can you include on your social media content calendar?

3. Where does your organization stand with your social media policy, and how can it be improved?

15
VOLUNTEERING IS A FAMILY AFFAIR

Being in the fire service is an honor and a blessing. All that good can become a curse if we allow the impacts and commitments of the fire department to destroy our marriage and family. Keep your family first and you can create a team that others want to join and stay a part of.

—Anne and Mike Gagliano,
authors of *Challenges of the Firefighter Marriage*

Importantly, within the fire service, *family* does not mean only spouse and children. Families are also tied to the situations an individual is in. The usage of the word "family" is as diverse as families themselves are. The National Volunteer Fire Council defines family in the fire service as, "The people who are related by birth, marriage, adoption, partnership, and those with whom you have a close relationship and consider family members."[1]

While the fire service often uses the idea of becoming part of the "fire service family" as a recruitment benefit, it is important to note, that for some individuals, the term "family" is not always associated with positive feelings. Some individuals may have experienced a family system with unhealthy interactions and behaviors between family members. Those with a dysfunctional family history may not be open to the idea of creating a new family within the fire service.

Family Engagement

Family engagement for successful volunteer firefighter retention was an emerging theme during my research. It is critical for fire service leaders to be proactive in involving the family from the beginning of the volunteer's fire service career. Volunteers who have left the industry frequently report conflict with balancing family life as a key factor for disengagement.² The following are reported impacts on family:

- Disruption of family routines due to volunteer deployment
- Time away from family for training and deployment
- Sleep deprivation impacts from volunteer deployment
- A shift in mood after returning from a call
- Behavioral health issues due to the repeated exposure to traumatic events
- The inability to reach loved ones during times of volunteer deployment
- Excessive stress and worry for family members during times of volunteer deployment

Family Stressors

Family members of firefighters have unique needs and strengths. These families can face diverse stressors during volunteer deployment. They are forced to adjust to their loved one being absent for inconsistent times during deployments. Family members must adjust to a break in normal family activities while processing the unknowns regarding the safety of their loved one(s).

The following are major stressors on families of volunteer firefighters:

- Feelings of loss from the absence of the firefighter
- Worry that their loved one won't return
- Decreased to no communication with the volunteer during deployment
- Parenting stress on the caregiver who is not deployed
- Occupational strain as a result of imbalanced volunteer-work-life balance

Historically, little emphasis has been placed on family-based interventions within the fire service. Volunteer firefighters also face challenges when reintegrating back into the family role after difficult calls. They may experience PTSD, rely on poor coping mechanisms, or just struggle to transition back into family life.

Stressful life circumstances can also negatively impact parenting and child outcomes.³ Consequently, fire service leaders need to create a family-friendly environment to reduce family conflict and stress. Reducing family conflict can also reduce disengagement among volunteers.

Family members of volunteer firefighters can also experience various behavioral health impacts because of the nature of their loved one's job. The constant stress, danger, and unpredictable schedules of firefighting can have significant negative behavioral effects on families. Potential behavioral health impacts on family members include the following:

- *Anxiety.* Constant fear and worry about the safety of their firefighter loved one, along with the inability to communicate with them, can lead to heightened anxiety levels among family members.
- *Depression.* The stress and emotional toll of firefighting can cause family members to experience depression, especially if they feel helpless or overwhelmed during times of volunteer deployment.
- *Emotional exhaustion.* Supporting a firefighter through challenging situations can be emotionally draining for family members because of the constant disruption to routine and exposure to traumatic events.
- *Sleep disturbances.* Irregular schedules, emergency calls during the night, and worrying while a loved one is deployed can all disrupt family members' sleep patterns.
- *Traumatic stress.* Being a sounding board for loved ones as they return from calls can lead to traumatic stress.
- *Vicarious traumatization.* As result of traumatic event, a family member may become overly involved or avoid their loved one, become hypervigilant and fear for their own safety, or have intrusive thoughts and images from stories they heard.
- *PTSD.* Family members may indirectly experience trauma through their loved one's stories and experiences, leading to symptoms of PTSD.
- *Social isolation.* The unpredictable nature of a firefighter's job can lead to missed family events and social gatherings, leading to feelings of isolation.
- *Coping with fear of loss.* The constant exposure to danger can force family members to cope with the fear of losing their firefighter loved one.
- *Relationship strain.* The demands of a firefighter's job may cause strain on family relationships due to time constraints and emotional stress.
- *Empathic strain.* Identifying with feelings and perspectives of loved ones, including pain and suffering, can make us disengage, turn away, or shut down in the face of the suffering of others.
- *Compassion fatigue.* An unhealthy work-life-volunteer balance that causes individuals to overextend themselves can lead to an extreme mental state characterized by burnout and secondary trauma.

It is essential for family members of firefighters to have access to support systems and resources to address these potential mental health impacts. Counseling, peer support groups, and open communication within the family can be beneficial in managing these challenges.

Becoming Family Focused

Fire departments that have become more family-focused have seen increased retention rates. Members of now-successful departments indicated that when they first joined, there was never a discussion about their spouse, ways for them to engage, or a focus on our family unit. Those departments have evolved to integrate family as a definitive part of the organization, which has led to higher retention.[4] They now have family members provide support during events, family-focused holiday parties and picnics; and make family at the station a part of the culture.

STRATEGY STOP

Create a culture for equitable family engagement within your organizations.

Creating a culture for equitable family engagement in the volunteer fire service involves promoting an inclusive and supportive environment that values the contributions of families of all types and recognizes the unique challenges fire service families face. The following tips will assist fire service leaders with achieving a culture that promotes family engagement:

- *Education and training*. Provide education and training opportunities for volunteers and their families to raise awareness about the importance of family engagement and address the specific challenges faced by volunteer firefighter families.
- *Communication*. Foster open and transparent communication between the fire department and firefighter families. A culture of open communication promotes an environment where families feel comfortable speaking openly. A culture of transparent communication promotes a safe space for topics that might not always be easy to bring forward. Ensure that families are regularly updated about critical department changes and any potential risks that can have a direct impact on their family.
- *Family support programs*. Establish support programs for firefighter families—for example, by offering resources for behavioral health, wellness, and financial advice.

- *Flexible scheduling.* Implement flexible scheduling options to accommodate the needs of firefighter families. This is critical for volunteers when it comes to honoring family commitments or dealing with emergencies at home.
- *Family-friendly events.* Organize family-friendly events within the fire department to strengthen the bond between firefighters and families.
- *Acknowledgment and appreciation.* Recognize the sacrifices and contributions of firefighter families through appreciation events, awards, or other forms of acknowledgment.
- *Family Advisory Councils.* Establish a formal Family Advisory Council, where family members have a platform to voice their concerns, process improvement ideas, and feedback. Providing a platform for family input will assist with future decision-making.
- *Work–life–volunteer balance.* Encourage and support work–life–volunteer balance for firefighters. This ensures that volunteers make family a priority and spend quality time with their families.
- *Cultural sensitivity training.* Provide cultural sensitivity training to volunteers to enhance understanding and respect for diverse family backgrounds and traditions.
- *Community involvement.* Engage firefighter families in community outreach programs to strengthen their sense of belonging and connection to the fire service.
- *Behavioral health support.* Offer behavioral health support services to firefighter families, recognizing the potential impact of the job on their mental well-being. The National Volunteer Fire Council (NVFC) offers members' families access to their behavioral health program at no change.
- *Zero tolerance for discrimination.* Enforce a zero-tolerance policy for discrimination or harassment based on family status within the fire service.

Ensure equitable family engagement by providing family engagement activities that foster opportunities for all and are not impacted by demographic, cultural, or other differences. Examples include the following:

- *Cultural sensitivity.* Certain cultures and religions have food restrictions, such as not being able to consume pork. Planning a pig roast may alienate some families. To overcome this, a department could still plan a pig roast but with food for everyone, to include vegetarian, vegan, and pork-free options. Be conscientious to communicate that the alternative options were separated to not touch meat or pork. It is also good practice to publish the menu in advance and invite feedback, to ensure that your selection accommodates everyone's beliefs

and traditions. In other words, both food diversity itself and communicating the food diversity are critical for event success.
- *Religious sensitivity.* Holiday parties are a great opportunity to bring families together at the fire department, but not everyone celebrates Christmas given their religious beliefs. December contains several holidays that reflect multiple cultural and religious traditions. Planning an end-of-the-year Christmas party may alienate some families. To overcome this, departments can instead hold an end-of-the-year family celebration to celebrate the entire team's contribution from the past year and to spark excitement for the new year. Encourage volunteers to dress for the party according to their holiday traditions; this allows everyone to embrace their personal views. Decorations can be tailored to a winter theme, or ask members to help provide decorations that represent their cultural traditions.
- *Socioeconomic demographic sensitivity.* Not all families can afford to contribute to or attend certain engagement activities. Ensure that optional family activities are low cost or free to ensure engagement. Avoid formal dress codes at parties, as it can restrict volunteers who cannot afford higher-end clothing. Be conscientious about the monetary and time value of what is being asked of families to contribute to events, potlucks, or other events.
- *Gender-responsive sensitivity.* Organizing an engagement event that responds well to the needs of diverse genders takes careful consideration, but this does not necessarily mean extra work or resources. Avoid asking event attendees to wear gender-specific fashions. Select a venue where all sexes and genders feel comfortable and that is accessible. For example, is the meeting being held in a club predominantly used by men that does not accommodate female needs? Does the venue have private restrooms and facilities for lactation? Is the event location free of any paraphernalia that may lead others to feel uncomfortable, such as posters with offensive language or photos? Is the language used in invitations gender neutral? Is the person organizing the event aware of gender issues? Do they have experience with organizing a gender-responsive event? Do they know who to ask for guidance?

By implementing these measures, the fire service can foster a culture that values equitable family engagement. This acknowledges the vital role families play in supporting firefighters and ensures the well-being of both firefighters and their loved ones.

STRATEGY STOP

Create joint training opportunities for volunteers and their family members.

Many outside organizations are looking for groups to present to. Fire departments can host joint training sessions with family members on topics that benefit both the firefighter and their families. Example topics include the following:

- Behavioral health
- Wellness
- Financial management
- Stress management
- Parenting
- Budgeting
- Retirement planning
- First aid and CPR

Case Study: Eating on a Budget

The Ohio State University was proud to partner with the Winona Fire Department volunteers to offer their volunteers and their families a series of classes to help them make healthier food choices on a budget.

—Yvette Graham, licensed independent social worker with supervision designation, The Ohio State University

Winona (OH) Fire Department partners with The Ohio State University

Family engagement is an important part of our fire department culture. In 2014 we partnered with The Ohio State University (OSU) to bring a wellness program to our firefighters and their family members (fig. 15–1). OSU provided free training to our volunteers and their families once a month over a period of 6 months. Classroom and hands-on sessions were provided and covered dietary quality, physical activity, food resource management (how to make healthy foods affordable), food safety, and food security. At the end of the training series, participants graduated and received a certificate from OSU. Of participants in the OSU program, 92% showed improvement in one or more nutrition practices, and 77% showed improvement in one or more food safety practices.

This program is known as the Expanded Food and Nutrition Education Program (EFNEP) and is funded by the U.S. Department of Agriculture's National Institute of Food and Agriculture. Land-grant universities conduct the EFNEP program in all states, and there is no cost to host this program.

Figure 15-1. OSU Program Specialist Danielle Massey teaching firefighters and their families about nutrition facts

It's easy to revert to an unhealthy eating habits. In particular, when unhealthy food is provided or bad habits are fortified by others, backsliding is likely. Getting the entire family involved in healthy eating can improve firefighter wellness. Inviting families to the fire department for training on health and wellness helps them to understand the importance of keeping their firefighters healthy to reduce the risk of cardiac events and other injuries.

To learn more about EFNEP and to find a program near you, visit the website of the National Institute of Food and Agriculture: https://www.nifa.usda.gov/grants/programs/capacity-grants/efnep/expanded-food-nutrition-education-program.

Thoughts to Consider

- What programs in your community can you partner with to offer educational programs for your firefighters and their families?

STRATEGY STOP

Create a family-friendly environment within your organizations for families to hang out.

Create a room or area within the department where families can hang out to play games, watch movies, and eat together. Creating a family-friendly environment in the fire department involves making conscious efforts to support volunteers and their families. Here are some strategies to achieve this:

- *Offer flexible scheduling.* Implement known flexible scheduling that allow volunteers to spend quality time with their families and attend important family activities without compromising their duties or feeling guilty for being absent from the department.
- *Provide family-friendly facilities.* Designate spaces in the fire department that are friendly toward families when they visit, such as a family room or a play area for children.
- *Host family events.* Encourage volunteers to involve their families in the department by organizing family-focused events and activities. These should be held regularly and scheduled as a family engagement plan (table 15–1).
- *Open communication.* Foster a culture of open and transparent communication between volunteers and their families by providing regular updates on schedules and key events.
- *Family visits.* Encourage and welcome family visits to the fire department to allow loved ones to get familiar with the environment, the people, and the work their loved one does.
- *Family training opportunities.* Offer family training programs, such as basic first aid or behavioral health workshops, to involve, educate, and care for families.
- *Family support resources.* Provide resources and information to support the families, such as behavioral health resources, financial advice, and child-care assistance.
- *Supportive leadership.* Ensure that fire department leadership actively supports family-friendly policies and initiatives and lead by example.
- *Family Advisory Councils.* Establish a formal Family Advisory Council, where family members can share feedback, suggestions, and concerns to influence decision-making within the fire department.
- *Child-care assistance.* Explore child-care assistance or referral options to help volunteer families to manage their responsibilities effectively.
- *Work-life-volunteer balance policy.* Implement and promote a work-life-volunteer balance policy that prioritizes the well-being of firefighters and their families.
- *Celebrating family milestones.* Celebrate important family milestones such as birthdays, anniversaries, graduations, and other special occasions to demonstrate the fire department's commitment to supporting families.
- *Parental support.* Offer support and resources for parents, such as parenting workshops or resources on balancing family life with demanding work or volunteer schedules.

Table 15-1. Sample family engagement plan

Date and Time	Activity	Event Notes
Saturday, July 1, 11 a.m.–1 p.m.	Firefighter Brunch Bunch	Families are invited to join firefighters for brunch. Pancakes, hash browns, and sausage links will be provided. Invitees are asked to bring their own beverage and favorite pancake topping to share.
Friday, July 7, 7 p.m.–9 p.m.	Friday-Night Flick: Goonies (rated PG)	Volunteers and families are invited to bring their own chairs and blankets as we turn the bay into a movie theater. Bring your favorite sweet snacks and beverages—and we will provide the popcorn.
Sunday, July 9, 1 p.m.–3 p.m.	Spikeless Sunday Sand Volleyball	Volunteers and family members are invited to play in a department tournament at Any Town Park. No spiking or overhand serves. Members of teams are required to be 50% female. Winning team gets bragging rights.
Friday, July 14, 6 p.m.–8 p.m.	Friday-Night Flick: Monsters University (rated G)	Volunteers and families are invited to bring their own chairs and blankets as we turn the bay into a movie theater. Bring your favorite sweet snacks and beverages—and we will provide the popcorn.
Saturday, July 15, 8 p.m.–?	Pizza & Game Palooza: Rocket League	Volunteers and families are invited to join together for a night of Rocket League. Bring your own controllers, beverages, and money to chip in for pizza.
Friday, July 21, 7 p.m.–9 p.m.	Friday-Night Flick: The Avengers (rated PG-13)	Volunteers and families are invited to bring their own chairs and blankets as we turn the bay into a movie theater. Bring your favorite sweet snacks and beverages—and we will provide the popcorn.
Saturday, July 22, 7 p.m.–?	Pizza & Board Game Palooza	Volunteers and families are invited to join together for a night of board games. Bring your own games, beverages, and money to chip in for pizza.
Saturday, July 29, 1 p.m.–3 p.m.	Saturday Swim	Volunteers and immediate family members are invited to join together for a free swim at the Any Town Community Pool. Pool passes will be provided in your department mailbox on July 27. Sign-up deadline is July 25.
Sunday, July 30, 3 p.m.–6 p.m.	Trailblazing Tribe Trot: Any Town Creek Trail	Volunteers and families are invited to join together for a group hiking adventure. Moderate difficulty including 2–4-mile options and some elevation change. The group will leave the fire department promptly at 3 p.m.

- *Respect for family time.* Have volunteers schedule "unavailable" time and encourage respect for family time during unavailable hours, ensuring that volunteers are not unnecessarily disturbed during their time with their loved ones.

By implementing these strategies, the fire department can create a welcoming and family-friendly environment that supports firefighters' personal lives while fostering a strong sense of camaraderie and commitment to their duties.

STRATEGY STOP

Create a family communication path within your organization.
Different individuals have varying perceptions of what communication should be like during volunteer training or deployment. It is critical to have conversations with volunteers about the need to establish communication goals with their loved ones, which will serve as a *family communication path*. Volunteers can be dispatched to an emergency when they are not with family members and may not have access to a phone, have time, or have a cell phone signal to let family know they responding to a call. The lack of being able to reach their first responder may lead to worry.

Establishing a family communication path can reduce communication frustration and increase relationship satisfaction. Communicating effectively with families of volunteer firefighters is crucial for maintaining transparency, building trust, and supporting their loved ones in the fire service. The following are tips for effective communication with families:

- *Regular updates.* Provide regular updates to firefighter families regarding training, work schedules, and any upcoming events or activities.
- *Open and transparent.* Be open and transparent in your communication, sharing both positive news and challenges faced by the volunteer firefighters.
- *Family meetings.* Organize family meetings periodically to discuss important matters, address concerns, and answer any questions.
- *Clear contact information.* Ensure that families have access to clear contact information for reaching out in case of emergencies or any urgent matters.
- *Respect boundaries.* Be respectful of family members' privacy and boundaries, especially during sensitive situations.

- *Family support resources.* Provide information about available resources and support services that can assist families in coping with the demands of their loved ones' volunteer firefighting role.
- *Training opportunities.* Offer family members the opportunity to attend training sessions or workshops related to firefighting, safety, or emergency preparedness.
- *Emergency protocols.* Clearly communicate emergency protocols to families so that they are aware of what to do in case of an emergency involving their loved one.
- *Acknowledgment and appreciation.* Show appreciation and acknowledgment to firefighter families for the support they provide to their loved ones and the community.
- *Social events.* Organize social events that involve both firefighters and their families, fostering a sense of community and camaraderie.
- *Understanding challenges.* Be empathetic and understanding of the challenges that firefighter families may face, and offer support as needed.
- *Two-way communication.* Encourage two-way communication, where families feel comfortable expressing their thoughts, concerns, and suggestions.
- *Multichannel communication.* Use various communication channels, such as emails, newsletters, social media, and phone calls, to reach families effectively.
- *Crisis communication plan.* Have a clear crisis communication plan in place to inform families during emergencies and critical situations.
- *Recognize family sacrifices.* Acknowledge and recognize the sacrifices made by firefighter families, expressing gratitude for their support.

Remember that effective communication is a two-way process. Encouraging open dialogue and actively listening to the needs of firefighter families will contribute to a stronger and more supportive relationship between the fire department and the families of volunteer firefighters.

STRATEGY STOP

Create a phone tree for families of volunteers.

A *phone tree* lists contacts in a branching structure that allows information to be quickly and efficiently disseminated through all members of the group. A phone tree for volunteer families can be a valuable communication tool during emergencies or critical situations. Here's how to develop a phone tree for first-responder families:

- *Identify a contact person.* Designate a reliable and responsible person within the fire department family network to act as the central point of contact for the phone tree.
- *Collect contact information.* Gather contact details for all volunteer families, including phone numbers (home, mobile, work), email addresses, and any other preferred communication methods.
- *Organize the tree.* Create a hierarchical list with the central contact person at the top. Divide the families into smaller groups, or *clusters*, and assign a group leader to each cluster.
- *Communicate the plan.* Inform all volunteer families about the phone tree and its purpose. Clearly explain the structure and how information will be relayed during emergencies.
- *Update contact information.* Regularly review and update the contact information in the phone tree to ensure accuracy and relevance.
- *Establish communication guidelines.* Set clear guidelines for when and how the phone tree should be activated. Determine who has the authority to initiate use of the tree (e.g., the incident commander or other department designee) and the preferred method of communication (phone call, text message, email, etc.).
- *Test the phone tree.* Conduct periodic drills to test the effectiveness of the phone tree.
- *Respect privacy and security.* Emphasize the importance of privacy and security. Instruct phone tree participants not to share the contact information outside of the designated phone tree.
- *Maintain communication etiquette.* Encourage respectful and concise communication within the phone tree, focusing on sharing essential information.
- *Provide training and awareness.* Hold training with all participants on how to use the phone tree effectively and raise awareness about its importance in emergency situations.
- *Enlist community support.* Collaborate with local authorities, first-responder agencies, and community organizations to strengthen the phone tree's effectiveness.

A phone tree is effective only if all users are committed to its proper use and maintenance. Regular communication and practice will ensure that volunteer families can rely on the phone tree when they need it most.

STRATEGY STOP

Create a communication fact sheet for families of volunteers. Communication is the basic building block of relationships. Good communication skills are essential for firefighter families.

Communication Fact Sheet for Families of First Responders

Emergency Contact Information: For emergencies involving your first responder, call 9-1-1 immediately.

Fire Department Nonemergency Contact: [Fire Department Phone Number]

[Fire Department Website]: For updates, safety information, and resources.

Phone Tree: We have a designated phone tree to relay important information during emergencies or critical situations. Please ensure your contact details are up-to-date with the [Contact Person's Name], the central point of contact.

Regular Updates: We will provide regular updates on your first responder's schedule, upcoming events, and safety protocols.

[Fire Department] Newsletter or Social Media: Stay connected and informed about fire department news and activities.

Training and Drills: Periodic drills and training sessions will be conducted to ensure everyone is prepared for emergency scenarios. Your participation in training events is encouraged to promote safety and awareness.

Community Support: We are committed to supporting each other as a community of first responder families. Reach out to your assigned group leader or [Contact Person's Name] for any concerns or questions.

Family Support Programs: We offer support programs to assist you with various needs, including mental health resources, financial advice, and child-care assistance. If you ever require any support, let us know and we'll guide you to the appropriate resources.

Work–Life Balance: We recognize the importance of work-life balance for first responders and their families. Efforts will be made to accommodate family events and responsibilities whenever possible.

Privacy and Security: Your contact information is stored securely and will only be used for communication within the fire department. Please refrain from sharing the phone tree contact details outside of its intended use.

Emergency Protocols: In case of an emergency involving your first responder, follow any instructions provided through the phone tree or official channels. Safety is our top priority, and we have established protocols to handle various emergency situations.

Open Communication: We encourage open and respectful communication among all first-responder families. Don't hesitate to express any concerns or make suggestions to improve our support system.

For additional information and resources, please visit our website at [Fire Department Website]. We are here to support you and your first responder. Together, we make a strong and resilient community.

STRATEGY STOP

Create a newsletter for families of volunteers.
A newsletter can be an effective tool to keep volunteers informed about the latest updates, news, and events within the organization. The following tips will help to make your newsletter informative and engaging for volunteers:

- *Regular schedule.* Send the newsletter on a consistent schedule (e.g., monthly or every other month) so that volunteers know when to expect it.
- *Clear and concise content.* Keep the newsletter content clear, concise, and relevant. Include essential information, updates, and announcements that directly impact volunteers.
- *Important dates and events.* Highlight upcoming events, training sessions, and important dates. Include such details as time, location, and any specific instructions.
- *Recognition and appreciation.* Use the newsletter to recognize and appreciate the efforts of volunteers. Highlight exceptional contributions and celebrate milestones.
- *Volunteer spotlight.* Feature individual volunteers and showcase their accomplishments. This helps to build a sense of community and connection among volunteers.
- *Training opportunities.* Provide information about upcoming training opportunities, workshops, and professional development sessions that may interest volunteers.
- *Success stories and impact.* Share success stories and the impact of the organization's work. Highlight how volunteers' efforts have made a difference in the community.

- *Visuals and multimedia.* Incorporate visuals, photos, and multimedia elements to make the newsletter visually appealing and engaging.
- *Interactive content.* Include interactive elements such as surveys, polls, or links to videos for volunteers to participate and provide feedback.
- *Important policies and procedures.* Share any updates or changes to organizational policies and procedures that may affect volunteers.
- *Contact information.* Provide contact information for key staff members and volunteer coordinators, making it easy for volunteers to reach out if they have questions or need assistance.
- *Mobile-friendly design.* Ensure the newsletter is mobile device friendly so that volunteers can access it on their smartphones or tablets.
- *Feedback and suggestions.* Encourage volunteers to provide feedback and suggestions for improving the organization and future newsletters.
- *Social media links.* Include links to the organization's social media platforms, encouraging volunteers to connect and stay updated through various channels.
- *Subscribe and unsubscribe options.* Provide clear options for volunteers to subscribe or unsubscribe from the newsletter based on their preferences.

Remember that the key to an effective newsletter is to make it valuable and relevant to volunteers. By keeping them informed, engaged, and appreciated, you can foster a strong sense of community and commitment among your volunteers.

Sample Newsletter Content

Dear Any Town Fire Department families,

Welcome to the latest edition of our Firefighter Family First Newsletter! We are excited to keep you informed about the happenings within the fire department and provide valuable resources to support you and your loved ones. As we continue to navigate through the year, we appreciate your unwavering support and dedication to our first responders. Together, we form a strong and resilient community.

Department News & Updates
Community Outreach Events. Join us in upcoming community outreach events, including fire safety workshops and school visits. We encourage your participation as we engage with our community to promote safety and awareness. Contact Joe Smith if you would like to get involved.

Safety Spotlight: Mental Health & Wellness
In this section, we emphasize the importance of mental health and wellness for our volunteers and their families. Check out some practical tips and resources to support overall well-being:

- *Recognizing Signs of Stress.* Learn how to recognize signs of stress in first responders and the importance of early intervention.
- *Coping Strategies for Families.* Discover effective coping strategies to help families deal with the unique challenges that come with a first responder's role.
- *Mental Health Resources.* Access a list of mental health resources available to all first responder families, offering support and guidance during challenging times.

Featured Family: Meet the [Last Name] Family
Get to know one of our incredible first responder families and their inspiring journey within the fire department. Their dedication and commitment to the community are truly commendable.

Upcoming Events & Celebrations
Stay updated with upcoming events, celebrations, and training opportunities. We encourage you to mark your calendars and get involved in these activities:

- *Firefighter Appreciation Day.* Save the date for our annual Firefighter Appreciation Day on [Date]. Join us in honoring our brave first responders and their families.
- *Family Picnic.* Let's come together for a day of fun, games, and bonding at our Family Picnic on [date]. It's an excellent opportunity to strengthen our community ties.
- *Family Support Corner.* Our Family Support Corner features resources and assistance available exclusively to first responder families.
- *Financial Planning Workshop.* Join our upcoming financial planning workshop tailored to first responder families' needs.

Community Spotlight: Recognizing Your Contributions
In this section, we showcase how first-responder families actively contribute to the betterment of our community. Share your stories of community involvement, and we'll feature them in future editions!

We hope you find this newsletter informative and inspiring. Our goal is to create a strong support system for all first responder families. Your feedback is invaluable to us, so please feel free to share your thoughts and suggestions with us.

Thank you for being an essential part of our fire department family. Stay safe and take care.

Onboarding for Family Members

Develop an *onboarding orientation*, at which family members are welcomed and familiarized with the fire department. Easing the minds of new recruits and their loved ones from the very beginning of the volunteer career can prevent feelings of isolation that can ultimately lead to volunteer turnover. The East Franklin Volunteer Fire Department in New Jersey prides itself on welcoming the *entire* family and has found this to be a successful tactic for retention.

Your onboarding orientation should not be about what not do, but should instead focus on welcoming the volunteer's family into the fire service family. The modules of the orientation and resources introduced should be tailored to ensure that each volunteer's entire family—and even their wider social network—feels informed and valued.

STRATEGY STOP

Develop a formal onboarding orientation for family members to ensure familiarization.

The following modules are suggested for onboarding:

- *Onboarding orientation for family members—introduction.* Welcome to [Your Fire Department Name]! We are delighted to have you join our extended family of firefighters. This onboarding training is designed to familiarize you with our fire department, its culture, and the support systems available for firefighter families. Your support is essential to the success of our firefighters, and we are committed to providing you with the resources you need to thrive as part of our community.
- *Module 1: Introduction to [Your Fire Department Name].* Get to know the history and mission of our fire department. Understand the roles and responsibilities of firefighters and the valuable contributions made by their families.
- *Module 2: Fire department operations.* Learn about the daily operations and functions of our fire department. Explore the various divisions and specialized units within the department.
- *Module 3: Firefighter safety and training.* Gain insights into firefighter training programs and the emphasis on safety protocols. Understand the importance of maintaining a safe environment for both firefighters and their families.
- *Module 4: Communication and resources.* Familiarize yourself with our communication channels and stay updated with department news and

events. Discover the family support resources available, including mental health services, financial advice, and child-care assistance.
- *Module 5: Emergency preparedness.* Learn about emergency procedures and what to do in case of a firefighter-related emergency. Be prepared to handle critical situations and support your firefighter during challenging times.
- *Module 6: Volunteer-work-life balance and family support.* Explore strategies for maintaining volunteer-work-life balance while supporting a firefighter in their demanding role. Discover the family-friendly initiatives and events organized by our fire department.
- *Module 7: Building a supportive community.* Understand the importance of building a supportive community among firefighter families. Learn about family events and activities that foster a sense of camaraderie and togetherness.
- *Module 8: Frequently asked questions.* Address common questions and concerns that families of firefighters may have. Provide access to additional resources for further information.

STRATEGY STOP

Develop a family onboarding checklist to ensure families are fully introduced to the fire department.

Sample Family Onboarding Checklist

Introduction to the Fire Department

- ☐ Fire department organizational overview
- ☐ Fire department organizational culture
- ☐ Fire department mission
- ☐ Fire department organizational chart
- ☐ Fire department family point of contacts

New Family Member Paperwork

- ☐ A copy of the NVFC's *Family Guide*
- ☐ Member handbook

Benefits Overview

- ☐ Health, life, and disability insurance
- ☐ Retirement benefits
- ☐ Educational assistance and training
- ☐ Employee assistance program
- ☐ Stipend procedures
- ☐ Incentive programs
- ☐ Discounts
- ☐ NVFC's First Responder Helpline (which is for family *and* volunteers)

Administrative Procedures

- ☐ Calendar of training and special events

Key Policy Review

- ☐ Antiharassment, antidiscrimination, and antiretaliation policy
- ☐ Dress code
- ☐ Personal conduct standards
- ☐ Progressive discipline
- ☐ Security
- ☐ Confidentiality
- ☐ Safety
- ☐ Emergency procedures
- ☐ Social media policy

Introductions and Tours

- ☐ Department staff and key personnel
- ☐ Township, city, or county personnel
- ☐ Tour of facility, including:
  » Restrooms
  » Showers
  » Computer room
  » Laundry room
  » Information and resource board
  » Member parking
  » Supplies
  » Dayroom
  » Kitchen
  » Gymnasium

STRATEGY STOP

Develop a formal mentoring network for family members to ensure family members feel a sense of connection with other family members.

A fire service *family mentor program* works with families to bring support, guidance, and instruction to help a volunteer's social network to navigate fire service life. Family mentors can help with advocacy and referral of behavioral health resources. Family mentors should be experienced family service members who are familiar with the fire department's mission, resources, benefits, administrative processes, and polices.

STRATEGY STOP

Develop a formal mentoring network for family members to ease their transition into fire service life.
The following are areas to include in the family mentorship program:

- *Introduction.* We are excited to introduce our Family Mentorship Program, designed to create a supportive network where experienced firefighter families can guide and mentor new or less experienced families. This program aims to foster a strong sense of community, provide valuable resources, and ease the transition for families joining our fire department family.

- *Program objectives*:
    » Create a welcoming and inclusive environment for firefighter families.
    » Offer guidance and support to new or less experienced families as they navigate their role within the fire department community.
    » Share knowledge, experiences, and resources to enhance the well-being and resilience of all firefighter families.
    » Strengthen camaraderie and foster a sense of togetherness among firefighter families.

- *Mentor and mentee selection.* Experienced firefighter families who have demonstrated commitment, understanding, and active involvement within the fire department will be invited to volunteer as mentors. New or less experienced firefighter families, upon joining the

department, will have the opportunity to express interest in becoming mentees.
- *Program activities and guidelines*:
  » *Orientation meeting.* At the program's start, all mentors and mentees will attend an orientation meeting to set expectations, clarify roles, and establish a comfortable mentorship relationship.
  » *Regular check-ins.* Mentors and mentees will have regular check-ins to discuss any challenges, concerns, or questions related to their role within the fire department community.
  » *Sharing experiences.* Mentors will share their experiences, insights, and coping strategies with mentees to help them to navigate the demands of being a firefighter family.
  » *Family support resources.* Mentors will introduce mentees to the family support resources available within the fire department, including mental health services, financial advice, and childcare assistance.
  » *Social events.* The mentorship program will organize social events where mentors, mentees, and their families can interact and build stronger connections.
  » *Training and workshops.* Mentors will encourage mentees to attend training sessions and workshops related to firefighting, safety, and emergency preparedness.
  » *Celebrating milestones.* The program will celebrate milestones and achievements of firefighter families, strengthening the sense of appreciation and support within the community.
- *Program duration.* The mentorship program will typically run for a duration of 6 months to 1 year, providing mentees with adequate support during their initial period within the fire department.
- *Benefits.* Access to a network of experienced firefighter families for guidance and support. Enhanced understanding of the firefighter family role, challenges, and coping mechanisms. Strengthened community ties and a sense of belonging within the fire department family. We believe that the Family Mentorship Program will be an invaluable resource for firefighter families, promoting a supportive and thriving community. Together, we create a stronger and more resilient family within our organization.

Battle Buddy System for Families

Develop a buddy-check system for family members by pairing individuals with a *battle buddy*. Battle buddies will watch out for one another in several ways—ensuring their match is making sound decisions, is linked to needed resources, and has someone to serve as a sounding board. The battle buddy system is also known as a *forced-best-friend system*, since the matched pair will actually become friends or will swap for a new person. Importantly, this system ensures that someone is always looking out for the family member—no one is left out.

A battle buddy's role encompasses the following:

- Be a friend to connect with during times of volunteer deployment.
- Provide the family member with at least one person to connect with when times become difficult.
- Check in at least once a month with their battle buddy (buddy checks can be as simple as a quick test message or a call).
- Routinely let your buddy know that you are OK by exchanging these simple words: "We are OK."

STRATEGY STOP

Develop a buddy-check system for family members by pairing individuals with a battle buddy.
The following outline can help you develop your battle buddy system.

Buddy-Check System for Family Members

Introduction:

At [Your Fire Department Name], we recognize the importance of providing support not only to our firefighters but also to their families. We are introducing our Battle Buddy System for Family Members, designed to create a network of mutual support, understanding, and camaraderie among firefighter families. The Battle Buddy System aims to strengthen the bonds within our fire department family and offer a helping hand during challenging times.

Objectives:

- Foster a supportive community among firefighter families by pairing them with a Battle Buddy from within the department.

- Encourage open communication and sharing of experiences between Battle Buddies to offer emotional and practical support.
- Promote resilience and well-being by providing families with a trusted ally they can turn to during both triumphs and struggles.

How the Battle Buddy System works:

- *Voluntary participation.* Family members of firefighters can voluntarily opt to participate in the Battle Buddy System during the onboarding process or at any time.
- *Buddy pairing.* Once a family member expresses interest in the program, they will be matched with a Battle Buddy from within the fire department.
- *Orientation meeting.* The paired Battle Buddies will attend an orientation meeting to get to know each other, set expectations, and establish a comfortable relationship.
- *Regular check-ins.* Battle Buddies will have regular check-ins to discuss any concerns, experiences, or challenges they may be facing.
- *Supportive listening.* Battle Buddies will actively listen and provide empathy, support, and encouragement during both positive and challenging times.
- *Resource sharing.* Battle Buddies will share helpful resources, tips, and coping strategies related to being a firefighter family.
- *Events and activities.* The Battle Buddy System will organize events and activities where paired buddies can participate and strengthen their connection.

Benefits:

- A dedicated support system for firefighter families within the fire department community
- A safe and confidential space for sharing experiences and seeking guidance
- A sense of belonging and camaraderie, knowing there is someone to lean on during times of need

Confidentiality and trust:

The Battle Buddy relationship is built on trust, respect, and confidentiality. All shared information will be treated with the utmost privacy, ensuring a safe space for open communication.

Join the Battle Buddy System:
To participate in the Battle Buddy System or for more information, please contact [Contact Person's Name] at [Contact Person's Email or Phone Number]. We encourage all firefighter families to take advantage of this opportunity for mutual support and connection.

Communication Tree for Family Members

A communication tree is a prearranged, defined system for disseminating information to a group of individuals. In a typical emergency phone tree, a designated call leader launches the phone tree by contacting the next people on the list. Those recipients then notify others in their cluster, and the process is repeated until everyone has been contacted. In large organizations, each person is usually responsible for calling at least two others.

STRATEGY STOP

Develop a communication tree for family members to use during incidents when information needs to be shared.
The following information will help with developing your plan for implementing a communication tree:

- A spreadsheet with no fewer than two contact methods per person, which can include work phone and home phone.
- Instructions specifying how notifications be made and any special procedures unique to that area.
- Sample scripts indicating what kinds of information the caller should share to accurately transfer information from person to person. Scripts should be concise so that the message can be delivered via phone call in less than 60 seconds. Scripts should incorporate necessary verbal cues including an introductory sentence about the situation, such as "This is a message from the Incident Commander...." Scripts should include a very brief statement of what has happened, what immediate actions should be taken, and where to get additional information.

Recruiting the Family

Additionally, expanding recruitment methods can lead to a new group of employees to retain. Recruiting family members of firefighters to help during fundraisers, community events, and critical incidents requires effective communication, involvement, and appreciation for their contributions.

STRATEGY STOP

Engage family members in active roles during fundraisers, community events, and critical incidents.

The following strategies can help you engage and recruit family members:

- *Ask—don't assume.* Don't assume every member of a firefighter's family will want to take an active role. They might be busy with work, serving other groups, or their family or just not have an interest. Active roles should be strictly volunteer; for example, no family member should be made to feel guilty for not helping with the annual fundraiser. Assigning roles against stated preferences or otherwise forcing engagement on family members may be counterproductive, leading to higher levels of dissatisfaction and less support to the volunteer.
- *Open communication.* Maintain open and transparent communication with firefighter families, informing them about upcoming fundraisers, community events, and critical incidents where their assistance is needed.
- *Family meetings.* Organize gatherings to discuss the importance of family involvement and how family members can support fundraisers and critical incidents.
- *Highlight contributions.* Recognize and appreciate the valuable role families play in the success of fundraisers and handling critical incidents. Highlight their contributions to boost morale and encourage further engagement.
- *Explain the impact.* Explain the positive impact that family involvement has had on the fire department's operations and the community it serves. Show how their efforts directly support the well-being of

their firefighter's loved ones and the success of the department's initiatives.
- *Engage in planning.* Involve family members in the planning process of fundraisers and critical incidents. Invite their input to make them feel part of the decision-making process.
- *Offer training and guidance.* Provide training and guidance to family members on how they can effectively assist during fundraisers and critical incidents. This may include information on fundraising techniques, crowd management, or supporting individuals during crisis situations.
- *Recognize time constraints.* Acknowledge the busy schedules of family members and offer flexible ways for them to contribute, such as online fundraising or remote support during critical incidents.
- *Express gratitude.* Regularly express gratitude for their involvement and the sacrifices they make to support their firefighter's loved ones. Showing appreciation reinforces their sense of purpose and belonging.
- *Create a family support network.* Establish a family support network within the fire department, where family members can connect, share experiences, and provide mutual assistance.
- *Share success stories.* Showcase success stories of families who have actively participated in fundraisers or critical incidents, emphasizing the positive outcomes of their involvement.
- *Provide resources.* Offer resources and guidelines to family members, such as a handbook on how to get involved, contact information for volunteer coordinators, and clear instructions on how they can contribute.
- *Celebrate achievements.* Celebrate the achievements of family members who have supported fundraisers and critical incidents, both big and small. Publicly recognize their efforts to motivate others to participate.
- *Engage social media.* Use social media to showcase family involvement during fundraisers and critical incidents, creating a sense of pride and encouragement among the wider fire department community.

Line-of-Duty Injuries and Deaths

Talking about the potential for line-of-duty injury and death can be difficult. Nevertheless, fire departments must have a conversation about the potential for injury or even death with family members and loved ones (fig. 15–2). This conversation should be approached with openness and trust, rather than with anxiety and fear.[5]

Figure 15-2. Memorial for a fallen firefighter

STRATEGY STOP

Addressing the topic of line-of-duty injuries and deaths to family members can be difficult.

Hold an educational session with volunteers and their loved ones to have a conversation about line-of-duty injuries and deaths. Conversations should include the following considerations:

- *Planning a funeral.* What special wishes surrounding the funeral or memorial service does the firefighter and family want? Is the honor guard requested at the service? Do you wish to have other departments present? Are there any religious or cultural needs?
- *Financial planning.* Discuss the life insurance policy offered by the department or through the NVFC (if a member), how to complete beneficiary documents, and how to access the policies in the event of an incident.
- *Preparing for serious injury.* Discuss how the family's lifestyle, finances, and mental wellness can be impacted if a debilitating injury occurs. What specific policies are provided or available regarding compensation and disability?
- *Outside help.* Within 24 hours of a firefighter's death, the fire department will notify the National Fallen Firefighters Foundation via the 24-hour line-of-duty death hotline at 866-736-5868. Inform the families that the Foundation can provide assistance with funeral planning and benefits filing, as well as much-needed support for the family and the department.

Review Questions

1. A fire service family mentor program works with families to bring _____ to help a volunteer's social network to navigate fire service life.

2. Forcing engagement on family members may lead to what?

3. Who does the fire department contact within 24 hours of a line of duty death?

Discussion Questions

1. What community organizations could provide trainings to your fire department members and their families?

2. What plan do you have for helping families in the event of line-of-duty injuries and deaths?

16

THE VALUE OF NONWAGE BENEFITS

The value of serving as a volunteer firefighter cannot be understated. When I returned from Vietnam, I immediately joined my volunteer fire department and took advantage of every training opportunity afforded. This led to a highly successful career in the fire service, returning after 28 years as a chief officer. Prepare, execute, and dream the dream!

—Bill Peters, battalion chief
Jersey City (NJ) Fire Department

Nonwage benefits are proven motivators for volunteer retention. However, these types of benefits are often overlooked by fire service leaders.

Nonwage benefits correlate with stakeholder theory by offering something of value to volunteers. Aligning nonwage benefits with the wants and needs of younger generations can also reduce turnover among volunteers. As indicated in table 16–1, nonwage benefits found valuable volunteer fire departments with high retention rates include uniforms and branded items, an attractive working environment, professional development opportunities, and camaraderie activities.

Table 16–1. Nonwage benefits at study site, for volunteer retention

Nonwage Benefit	Frequency of Times Mentioned	% of Times Mentioned by Participants
Uniforms and branded items	37	36%
Attractive working environment	11	11%
Professional development opportunities	39	38%
Camaraderie activities	16	16%

A Path to a Future Career

Some of the most influential firefighters in the country got their start in the volunteer fire service. They leveraged their volunteer service for competitive advantage. Their success underscores that volunteer firefighting can be a stepping-stone to a career position.

The volunteer service often perceives that there is a threat of losing members to the career side. However, fire service leaders need to rethink this position and instead market volunteer firefighting as a career path. Volunteer departments can feed into career departments and health care professions.

The path to becoming a career firefighter may vary depending on the specific requirements of different fire departments and regions. Regardless, volunteering as a firefighter undoubtedly provides a solid foundation and valuable experience for those aspiring to a full-time firefighting career.

The following are ways volunteering as a firefighter can lead to a job as a career firefighter:

- *Experience and training.* Volunteer firefighting provides valuable hands-on experience and training in firefighting techniques, emergency response, and equipment operation. This experience is highly beneficial when applying for a career firefighting position.
- *Networking and references.* Volunteering allows you to build connections within the firefighting community. Networking with career firefighters and fire department personnel can lead to job opportunities and positive references.
- *Proven dedication.* Volunteering shows your dedication to the firefighting profession. It demonstrates your commitment to public safety and your willingness to serve the community, making you a more attractive candidate for career firefighter positions.
- *Enhanced skills.* As a volunteer firefighter, you develop a diverse skill set, including teamwork, leadership, problem-solving, and critical decision-making. These valuable skills are transferable to any firefighting role.
- *Understanding of fire service culture.* Volunteering provides insight into the fire service culture, operations, and organizational structure. Understanding the dynamics of a fire department can help you to adapt quickly to a career firefighting role.
- *Physical fitness.* Firefighting requires a high level of physical fitness. Volunteering helps you to maintain and improve your fitness

level, making you more prepared for the physical demands of a career as a firefighter.
- *Proving yourself as a candidate.* Volunteer firefighting allows you to demonstrate your passion and commitment to firefighting as a profession. When career firefighter positions become available, your experience as a volunteer can make you stand out as a strong candidate.
- *Earning certifications.* Many volunteer firefighters have the opportunity to earn certifications and qualifications related to firefighting and emergency response. These certifications can enhance your resume and increase your chances of securing a career firefighting job.
- *Increased job opportunities.* Many fire departments prefer hiring individuals with firefighting experience, which volunteer firefighting can provide. Your volunteer experience may give you an advantage over other candidates without prior firefighting experience.
- *Proving your skills in real situations.* Volunteering exposes you to real-life emergency situations, where you apply your firefighting skills and knowledge. Having experience in real-life scenarios can be valuable in career firefighter interviews and assessments.

Uniforms and Department-Branded Items

The brand of a fire department is important for retaining volunteers. The organization's name and its internal stakeholders all make up the brand. In fact, wearing a uniform branded to the fire department is one motivating factor for volunteers in emergency services.

Fostering a sense of belonging and personal pride within the fire department contributes to a collective pride toward the organization. Moreover, doing so represents an effective strategy for retention. To develop these feelings, give volunteers branded shirts and other department paraphernalia with the organization's logo. Providing both class A and B uniforms is a key strategy.[1]

STRATEGY STOP

Provide uniforms and department-branded items for your members.
Uniforms and shirts can be earned by volunteers through career achievements (table 16–2). These rewards also signify that the volunteer worked hard to be

Table 16–2. Examples of branded items for distribution

Branded Item	Milestone to Achieve Item	Presentation
Fire department T-shirt	Becoming sworn in as a member of the fire department	Issued during the welcome message at the first monthly meeting the new volunteer attends
Class B uniform	Receiving their first certification, such as Firefigher I or EMT	Issued just before graduation—to be worn for Graduation/class pictures
Fire department vehicle sticker	Completion of the online or in-person Traffic Incident Management course	Issued at the monthly meeting that falls after completion of the course
Class A uniform	After 1 year of service and being removed from probation	Issued at the monthly meeting that falls after the member's 1-year anniversary with the organization
Fire department quilted jacket	After 2 years of service	Issued at the monthly meeting that falls after the member's 2-year anniversary with the organization

part of the team. The uniform reflects a sense of belonging and community for the volunteer.

Offer an Attractive Work Environment

Providing volunteer firefighters with a fun and attractive work environment has been proven to increase volunteer satisfaction and engage younger generations.[2] For example, to retain younger generations, successful organizations have added on-site pool tables, volleyball pits, and rock-climbing walls and hosted sports leagues for members. However, one limitation with such fun-focused volunteer retention measures is that older generations with limited financial resources obtain higher motivation from stipends than from environmental incentives.

STRATEGY STOP

Create an attractive work environment for your members.
Having an environment that encourages volunteers to hang out has been found to be a successful method for retention. The following are ideas to make your fire department more attractive:

- Create a dayroom with ample seating for hanging out
- Provide video game systems and an area for group play
- Install large-screen televisions with movie players
- Allow personal use of a washer and dryer by volunteers
- Supply refreshments such as soda, water, and popcorn
- Set up a quiet study area with Wi-Fi (wireless Ethernet), printing, and docking capabilities
- Offer a bunk room for overnight stay by members who live outside of the coverage area
- Create a workout area with diverse equipment

Investment in Professional Development

Organizations that offer professional development and training have higher rates of job satisfaction and retention among volunteers. The fire service is an industry filled with opportunities for growth. Offering free education can have a positive impact on volunteer engagement. In particular, millennial volunteers have a greater interest in meaningful roles over salary and are keen on professional development for career progress. One fire service leader representative I interviewed reported that training provided by the organization allowed their volunteers to become leaders in other organizations, and another said that training kept morale high in their organization.

The fire service, like any other business, is in need of diverse skill sets to stay fully operational. Instead of outsourcing services, fire service organizations can invest in their own people. This meets not only the needs of the fire department but the needs of the volunteers. It gives them skills that they can add to their resumes. Adult career centers and community colleges are filled with professional development courses and certifications that can be used to achieve this goal.

STRATEGY STOP

Offer professional development opportunities for your members. Professional development opportunities do not have to be fire service focused—they can include anything that will benefit the department.

For example, having a website is key for visibility among stakeholders; moreover, online form submission is the preferred way for potential members to complete the volunteer application. As technological needs, many departments outsource webpage development and maintenance. However, fire service organizations can eliminate the monthly cost of a webmaster by providing training to members on how to use web platforms. This is a win-win: The department gains in-house webmasters, and the members being trained gain professional development that may be beneficial in the job world.

Who do you pay to repair and conduct daily maintenance of your vehicles? Would paying for a member or two to become certified mechanics reduce the costs of routine maintenance?

Are you engaged in large community events, fundraisers or acquisition projects? Would sending someone to learn how to become a project manager increase the efficiency in how you operate and reduce costs of big projects? The most successful organizations in the world use project managers, so why can't you?

Who is developing your recruitment and marketing materials? Would sending your most creative volunteer to a course in Photoshop and graphic design allow you to create professional-looking web and print materials internally, instead of paying a marketing firm?

The list of opportunities is endless, and our younger generations have a greater interest in meaningful roles over financial gain. Offering professional development tied to a role within your organization increases the feeling of organizational ownership and the volunteer's desire to engage.

STRATEGY STOP

Provide tuition reimbursement for your members.
With the rising cost of both college and trade school, the younger generation welcomes ways to avoid student loan debt. The average cost of college in the United States is $36,436 per student per year, including books, supplies, and housing.[3]

Fire departments that have implemented tuition benefits in return for service successfully used this nonwage benefit to attract volunteers. For example, Maryland State Firefighters Association (MSFA) used SAFER funding to offer this nonwage benefit to their membership. The MSFA listed the following:

> Purpose: It shall be the policy of the MSFA to maximize the retention of qualified active volunteer fire and rescue personnel through the use of a SAFER-funded College Tuition Reimbursement Program for any active and in good standing member or a member of their immediate family.

Details: Applicants MUST be current active members in good standing within their fire department. Applicant must show cause that they cannot afford the tuition for their child or themselves without having to take on a second or greater job that would reduce or eliminate the time that they currently spend volunteering in the community. Applicants or applicants' family member must be a high school graduate or equivalent.

Tuition Reimbursement is for any field of study at an accredited institution of higher learning. Students may only apply for two semesters in a single reporting period. A maximum of $4,000 will be reimbursed to each qualified and approved recipient, regardless of total funds expended by the student.

Should there be insufficient applications, the SAFER Grant Project Manager reserves the right to provide additional reimbursement to qualified recipients on a case by case basis, based upon participation and level of school activity.

The decision on funding Tuition Reimbursements shall be made within ten (10) days following the submission window for this application period. The amount of the award may not exceed the actual cost of tuition and fees charged for courses.

When accepting money via the Tuition Reimbursement Program, recipients must sign a legal contract committing to the provision that the applicant actively provides services for a period of one year following the award. In doing so, the recipient acknowledges that the failure to do so will result in a legal obligation to remit the exact amount of the reimbursement to the MSFA within ninety (90) days of the end of the calendar year in which the recipient failed to meet stated requirements.

This is a Tuition Reimbursement Program. Qualified recipients will only receive monies after successful payment of funds of the semester for which funding is being requested and with proof [of] current enrollment.

Camaraderie Activities

Making fun a part of the work environment is another method for volunteer commitment. Millennials seek a fun work environment, and building camaraderie through activities can realize this goal. Organization-sponsored activities allow volunteers the opportunity to build personal relationships with one another and create a higher level of camaraderie.

Activities that promote camaraderie may drive stakeholder engagement. Consequently, organizations that offer such activities indicate having higher levels of productivity and commitment. Organizational events and games keep volunteers inspired. Winning games or tournaments may also result in long-term pride for a volunteer.

Camaraderie activities have been ranked as one of the best aspects in the workplace. One fire service leader reported that picnics for different holidays, social gatherings, and dinners are camaraderie activities that kept their volunteer firefighters interested.[4]

STRATEGY STOP

Offer camaraderie activities for your members.
Camaraderie activities are essential for building strong bonds and fostering a sense of unity among volunteer firefighters. These activities promote team spirit, trust, and cooperation, which are vital for an effective firefighting team. The following is a list of camaraderie activities for volunteer firefighters:

- *Team-building exercises.* Plan team-building activities such as ropes courses, obstacle courses, or escape rooms. These activities encourage collaboration, problem-solving, and communication among team members.
- *Firefighter games.* Organize friendly firefighter games or competitions, such as tug-of-war, relay races, or a firefighting-themed scavenger hunt.
- *Community events.* Participate together in community events such as building a house for Habitat for Humanity, parades, fundraisers, or charity runs. These activities allow firefighters to bond with the community and each other.
- *Sharing meals.* Host cookouts or barbeques at the fire station or a local park. Sharing a meal together fosters a relaxed and social atmosphere.

- *Movie or game nights.* Arrange movie nights or game nights at the fire station. Watching a movie or playing games together can be a fun and enjoyable way to bond.
- *Group sports outings.* Attend sporting events together as a group. Whether it's a local game or a professional match, a sporting event provides an opportunity for firefighters to enjoy recreational time together.
- *Sporting activities.* Sign up together for sand volleyball, a golf tournament, a softball leagues, a bowling team, or any other nonprofessional sports groups in the community.
- *Volunteer projects.* Collaborate on volunteer projects within the community. This could involve helping with a community cleanup, assisting elderly residents, or supporting local charities.
- *Camping or retreats.* Organize camping trips or retreats, providing an opportunity for firefighters to unwind and bond in a different setting.
- *Celebrations and milestones.* Celebrate birthdays, work anniversaries, and other milestones among team members. Recognizing individual achievements strengthens the team's morale.
- *Training competitions.* Host friendly training competitions where teams compete in scenarios using various firefighting skills. This can be both educational and entertaining.
- *Wellness programs.* Implement wellness programs that focus on physical and mental health. Group activities such as yoga or fitness classes can promote well-being and camaraderie.
- *Holiday parties.* Host holiday parties or themed events during festive seasons, encouraging a joyful and festive atmosphere among the firefighters.
- *Sharing personal stories.* Create opportunities for firefighters to share personal stories or experiences. This can deepen understanding and empathy among team members.

Remember to consider the preferences and interests of your volunteer firefighters when planning these activities. The goal is to create a positive and supportive environment where firefighters can bond, collaborate, and feel a strong sense of camaraderie within the team.

Review Questions

1. Aligning nonwage benefits with the wants and needs of younger generations organization can achieve what?

2. What is one limitation to creating a fun environment for volunteer retention?

3. Professional development opportunities do not have to be fire service focused, but can also include things that _____.

Discussion Questions

1. How can you make your department a more attractive work environment?

2. What professional skills does your department need that you could send someone to training?

17
RECOGNIZING VOLUNTEER SUCCESS

We have an annual installation of officers dinner in January and what we do there is we recognize the top 10 responders of the company, those that come to the most fire calls for the previous year. When they go above and beyond and achieve very good things here we recognize them at this installation dinner, whether it be the Top 10 Responder award, we have Rookie of the Year award, Firefighter of the Year award, Fire Officer of the Year award, things of that nature, so that people don't go unrecognized for their good doings here at the volunteer fire department.[1]

—East Franklin (NJ) Fire Department

Recognizing volunteer firefighter success is essential for boosting morale, fostering a positive team culture, and motivating volunteers to continue their dedication. Accommodating volunteer success relates to stakeholder theory, and offering opportunities for volunteer success in the workplace is a key retention strategy. Leaders who lack effective strategies to increase volunteer morale face a decrease in productivity, sustainability, and profitability; increasing absenteeism and turnover; and a negative financial impact.[2]

Create Recognition Programs

Recognition programs create value for volunteers and promote engagement with department values. Recognizing and celebrating volunteer accomplishments and special occasions leads to a shared sense of responsibility among volunteers and an increase in job satisfaction.

Implementing a volunteer recognition program can assist with the following:

- *Acknowledgment and appreciation.* Recognition programs provide a formal way to acknowledge the efforts and contributions of

volunteers. Feeling valued and appreciated encourages volunteers to remain committed to their roles.
- *Morale and motivation.* Being recognized for their hard work boosts volunteers' morale and contributes to their motivation. It reinforces their sense of purpose and satisfaction, leading to increased enthusiasm and dedication.
- *Retention and engagement.* Volunteer recognition programs help improve volunteer retention rates. Volunteers who feel appreciated are more likely to stay committed to the organization and continue their service.
- *Positive organizational culture.* A culture of recognition fosters a positive environment within the organization. Volunteers are more likely to enjoy their volunteering experience and feel part of a supportive community.
- *Peer recognition.* Recognition programs often involve peer-to-peer recognition, where volunteers acknowledge each other's efforts. This encourages teamwork and camaraderie among volunteers.
- *Inspiring others to get involved.* Publicly recognizing volunteers' achievements can inspire others to join and contribute to the organization. Positive recognition acts as an advertisement for the organization's volunteer opportunities.
- *Sense of pride.* Recognizing volunteers' efforts instills a sense of pride in their work and accomplishments. This sense of pride encourages volunteers to continue making a difference in their community.
- *Improved performance.* Volunteers who feel valued are more likely to perform at their best. Recognition can lead to increased commitment and productivity in their roles.
- *Enhanced community relations.* Recognizing volunteers' contributions publicly can improve the organization's reputation and strengthen community relations.
- *Support for volunteer recruitment.* Volunteer recognition can be highlighted in recruitment efforts, attracting potential volunteers who seek appreciation and acknowledgment for their contributions.
- *Feedback mechanism.* Recognition programs can serve as a feedback mechanism. When recognizing volunteers, leaders can communicate specific contributions that align with the organization's goals.
- *Celebrating diversity.* Recognition programs can celebrate the diverse range of skills and experiences that volunteers bring to the organization. This highlights the inclusive nature of the volunteer community.

- *Long-term commitment.* Long-serving volunteers often dedicate countless hours to the organization. Recognition programs show gratitude for and sustain this ongoing commitment, encouraging volunteers to continue their service.

By appreciating and acknowledging the contributions of volunteers, organizations foster a positive environment, encourage retention, and attract new volunteers to support their mission and goals.

STRATEGY STOP

Implement recognition activities for your members.
The following are ways to acknowledge and celebrate the success of your volunteers:

- *Public appreciation.* Recognize volunteers during public events, community gatherings, or town meetings to showcase their contributions and commitment to the community.
- *Certificates and awards.* Present certificates of appreciation or special awards to volunteers for outstanding service, leadership, or exceptional efforts.
- *Verbal acknowledgment.* Publicly acknowledge and praise volunteer firefighters, with their permission, during team meetings, training sessions, or fire department events.
- *Social media shoutouts.* Highlight volunteer success stories, with their permission, on social media platforms, showcasing their achievements and contributions to the community.
- *Wall of Fame.* Create a Wall of Fame or similar recognition board in the firehouse, where successful volunteers are featured for their outstanding service.
- *Peer recognition.* Encourage a peer-recognition program, where members can nominate their fellow volunteers for exceptional efforts or teamwork.
- *Media coverage.* Seek media coverage for significant accomplishments, community projects, or successful emergency responses involving volunteer firefighters.
- *Volunteer spotlights.* Feature regular volunteer spotlights on your fire department's website or newsletter, showcasing individual volunteers, with their permission.
- *Thank-you notes.* Send personalized thank-you notes or emails to volunteer firefighters, expressing gratitude for their dedication and success.

- *Training opportunities.* Offer opportunities for specialized training or certifications as a reward for successful achievements.
- *Service milestones.* Celebrate significant service milestones (e.g., years of service) with special events or gifts to acknowledge long-term commitment.
- *Volunteer of the [Month/Year].* Establish a Volunteer of the Month or Volunteer of the Year recognition program, through which outstanding volunteers are honored for their contributions.
- *Special assignments.* Assign volunteers to special roles or tasks as a way to recognize their skills and expertise.
- *Volunteer appreciation events.* Host volunteer appreciation events, such as dinners or gatherings, to celebrate the collective success of the entire volunteer firefighting team.
- *Family involvement.* Involve the families of volunteer firefighters in recognition events to show appreciation for their support and sacrifices.

Remember that recognition should be genuine, timely, and tailored to the individual volunteer's efforts. Each volunteer firefighter brings unique strengths to the team. Recognizing their successes fosters a sense of pride and commitment to the fire department's mission.

Establishing a Length-of-Service Award Program

Length-of-service award programs (LOSAPs) are another way to recognize the contributions made by volunteers in the fire service. A LOSAP is a county- or municipal-funded program that provides a cash award monthly or on another regular basis (varies by location) for members who have served a specific number of years with a volunteer fire, rescue, or emergency medical services (EMS) department. LOSAP is "a system established to provide tax-deferred income benefits to active volunteer members of an emergency service organization."[3] Under this system, contributions are deposited into a 457(e)(11) plan, similar to plans for government employees. The volunteer or governing body then directs the investment of funds.

Importantly, LOSAP availability and rules vary by state. For example, in Ohio, volunteers do not have an established LOSAP program.

During my study, the LOSAP was mentioned by 90% of participants as an effective retention strategy. They reported activities that counted as active service, including meetings, community service, fire calls, training, truck maintenance, cleaning, and duties related to an officer position.

State and National Awards

Integrate public-facing recognition mechanisms into your strategy to recognize volunteers. In addition to having a positive impact on the volunteer being recognized, state and national awards often provide good publicity for the fire department.

STRATEGY STOP

Nominate your volunteers for awards at all levels: local, state, and national.

To nominate volunteers for state and federal awards, you would typically follow these steps:

- *Research eligibility.* Look into specific state and federal awards available for volunteers. Check eligibility criteria, nomination deadlines, and required documentation.
- *Select deserving volunteers.* Identify volunteers who have made significant contributions and demonstrated exceptional dedication to the organization or the community.
- *Gather information.* Collect detailed information about each nominee, including their volunteer work, achievements, impact on the community, and any relevant supporting documents.
- *Personalize nomination letters.* Write compelling nomination letters for each volunteer. Highlight their accomplishments and unique qualities, as well as the ways in which they have made a difference.
- *Submit nominations.* Nomination letters and any required documentation must be submitted by the specified deadline. Follow the application instructions provided by the awarding organization.
- *Notify nominees.* Inform the nominated volunteers about their nomination and let them know about the process and timeline for award selection.
- *Prepare for interviews or verification.* Depending on the award process, some nominees may be required to participate in interviews or undergo a verification process.

- *Celebrate and share.* Regardless of whether a volunteer receives an award, celebrate their efforts and contributions. Don't forget to share the "wins" of their stories and impact with the community and the organization.

Remember that the process for nominating volunteers for state and federal awards may vary depending on the specific award and the organization presenting it. It's essential to carefully review the guidelines and requirements for each award for the highest likelihood of a successful nomination.

Consider the following three national awards:

- President's Volunteer Service Award
- National Volunteer Fire Council awards
- Congressional Fire Service Institute awards

Case Study: Submit Your Success Stories

Congressional Fire Service Institute/Masimo Excellence in Fire Service-Based EMS Award Program

When I joined the Winona (OH) Fire Department, we provided fire suppression only. We quickly recognized a need for EMS in our rural area. I was tasked with implementing an EMS program and served as the department's first EMS chief officer. We are a small department that responds to a 33-square-mile rural community of approximately 4,700 residents.

Our EMS program became so successful that we submitted for and received the Congressional Fire Services Institute Excellence in Fire Service-Based EMS Award in 2019. We never thought that a small community like ours would be recognized on a national level. After all, small fire departments often feel like they can't compete with big city fire departments. Our selection shows perfectly how untrue that mind-set is; submitting local success stories for national awards can lead to a big win.

The Excellence in Fire Service-Based EMS Award led to many great outcomes. It was great for volunteer morale, we received a high level of community publicity, and we used it as leverage when asking for donations.

Thoughts to Consider

- What success stories does your department have to share?
- What local, state, and federal awards are you aware of?

Review Questions

1. Recognizing and celebrating volunteer accomplishments and special occasions leads to a shared _____ among volunteers and an increase in job satisfaction.

2. A LOSAP provides _____ to active volunteer members of an emergency service organization.

Discussion Questions

1. How can recognize your volunteers locally?

2. How can you recognize your volunteers at a state or national level?

18

TRACKING YOUR PROGRESS

Tracking the progress of recruitment and retention efforts is crucial for evaluating their effectiveness—and to make necessary adjustments. Many fire service organizations fail to or are unaware of how to implement a process to track their progress.

Tracking Recruitment Efforts

By tracking the progress of your recruitment plan, you can identify strengths and weaknesses, make data-driven decisions, and optimize your efforts to attract dedicated and diverse volunteer firefighters to your team.

STRATEGY STOP

Track your recruitment efforts.
The following steps can help you to track your progress:

- *Establishing measurable goals.* Define clear and measurable recruitment goals, such as the number of new volunteers to be recruited within a specific time frame or increasing the diversity of the volunteer pool.
- *Data collection.* Create a system for collecting relevant data throughout the recruitment process. This may include the number of applicants, recruitment sources, demographics, and application completion rates.
- *Tracking recruitment sources.* Keep track of where potential volunteers hear about your recruitment efforts. This could include social media, local events, community fairs, word of mouth, or specific advertising campaigns.
- *Application completion rate.* Monitor the completion rate of volunteer firefighter applications to identify any potential barriers or areas for improvement in the application process.

- *Engagement and response time.* Measure the responsiveness of your recruitment team to inquiries and applications. Prompt and personalized responses can positively impact the recruitment process.
- *Retention rate.* Track the retention rate of newly recruited volunteers to understand how successful your recruitment efforts are at attracting committed individuals.
- *Assessment of outreach efforts.* Evaluate the effectiveness of different outreach efforts, such as recruitment events, informational sessions, or targeted advertising, to determine which methods yield the best results.
- *Demographic analysis.* Analyze the demographics of the recruited volunteers to ensure inclusivity and diversity in your fire department.
- *Feedback collection.* Gather feedback from both successful and unsuccessful applicants about their recruitment experience to identify areas for improvement.
- *Comparison with previous periods.* Compare the current recruitment plan's progress with past recruitment periods to identify trends and improvements.
- *Regular review meetings.* Conduct regular review meetings with the recruitment team to discuss progress, challenges, and potential adjustments to the plan.
- *Recruitment software.* Consider using recruitment software or tools that can automate data collection and provide real-time insights into your recruitment progress.
- *Benchmarking.* Compare and contrast your recruitment plan against best practices in volunteer firefighter recruitment to identify areas for enhancement.
- *Celebrating milestones.* Celebrate successes throughout implementation of your recruitment plan, such as reaching a specific number of new recruits or achieving diversity goals.
- *Adapting and improving.* Based on the tracked data and insights, adapt and improve your recruitment plan continuously to optimize its effectiveness.

Tracking Retention Efforts

Tracking the progress of a retention plan for volunteer firefighters is essential to assess its effectiveness and make necessary adjustments to improve firefighter retention.

STRATEGY STOP

Track retention efforts.
Regularly track the progress of your retention plan to identify strengths and weaknesses, measure the plan's impact, and make data-driven decisions that will enhance firefighter retention in your volunteer fire department. The following strategies will help you to track the progress of your retention plan:

- *Establishing measurable goals.* Define clear and measurable retention goals, such as reducing attrition rates, increasing volunteer satisfaction, or improving the length of service.
- *Data collection.* Establish a system for collecting relevant retention data, including the number of volunteer firefighters, their length of service, reasons for leaving, and feedback on their experiences.
- *Exit interviews.* Conduct exit interviews with departing volunteer firefighters to understand their reasons for leaving and gather insights to improve retention efforts.
- *Volunteer feedback.* Regularly survey current volunteer firefighters to gather feedback on their experiences, satisfaction levels, and suggestions for improvements.
- *Retention rate.* Calculate the retention rate by comparing the number of retained volunteer firefighters with the total number at the beginning of a specific time period.
- *Demographic analysis.* Analyze the demographics of retained and departed volunteer firefighters to identify any patterns or disparities that may influence retention.
- *Tracking retention initiatives.* Monitor the implementation and impact of specific retention initiatives introduced as part of the retention plan.
- *Training and development participation.* Track participation rates in training and development programs among volunteer firefighters to assess their engagement and potential influence on retention.
- *Performance evaluations.* Monitor performance evaluations and recognition practices to identify potential links between recognition and retention.

- *Retention committee review.* Hold regular meetings with the retention committee to review progress, identify challenges, and discuss potential adjustments to the plan.
- *Benchmarking.* Compare and contrast your retention plan against industry best practices or compare with other fire departments to assess your efforts in context.
- *Retention interviews.* Conduct retention interviews with volunteer firefighters who have been with the department for an extended period to gain insights into what factors contribute to their long-term commitment.
- *Celebrating milestones.* Recognize and celebrate the service milestones of long-serving volunteer firefighters to foster a culture of appreciation and loyalty.
- *Feedback from leadership.* Seek feedback from department leadership and officers on the effectiveness of retention strategies and any suggestions for improvement.
- *Continuous improvement.* Use the data and insights gathered to continuously improve and refine your retention plan, adapting it to address emerging challenges and opportunities.

Tracking Volunteer Satisfaction

Tracking the satisfaction of volunteer firefighters is crucial for understanding their experiences, identifying areas for improvement, and ensuring their continued engagement and commitment.

STRATEGY STOP

Track the satisfaction of your volunteers.
By tracking the satisfaction of volunteer firefighters, you can proactively address issues, enhance their experiences, and create a positive and rewarding environment that encourages continued commitment and engagement. The following strategies can help you to track the satisfaction of volunteer firefighters:

- *Satisfaction survey.* Create a comprehensive satisfaction survey tailored to the specific needs and concerns of volunteer firefighters. Include

questions related to their experiences, challenges, training, work environment, and overall satisfaction.
- *Anonymity and confidentiality.* Assure volunteer firefighters that their responses will be anonymous and confidential, encouraging honest feedback.
- *Survey distribution.* Distribute the satisfaction survey through various channels, such as email, online platforms, or printed forms. Ensure that it is easily accessible and user-friendly.
- *Regular survey schedule.* Establish a regular schedule for conducting the satisfaction survey (e.g., annually or biannually) to track changes in satisfaction over time.
- *Include open-ended questions.* In addition to multiple-choice questions, include open-ended questions to allow volunteer firefighters to express their opinions and provide specific feedback.
- *Aggregate and analyze data.* Gather and analyze the survey data to identify trends, patterns, and areas of concern that may affect volunteer satisfaction.
- *Benchmarking.* Compare and contrast the survey results with industry data or previous survey data to gain insights into any improvements or declines in satisfaction levels.
- *Data segmentation.* To understand if satisfaction levels vary among different groups, segment the data by demographics, length of service, and firefighting roles.
- *Feedback from leadership.* Seek feedback from department leadership and officers on their observations regarding volunteer firefighter satisfaction and possible areas for improvement.
- *Consideration of retention data.* Correlate the satisfaction survey data with retention data to assess the impact of satisfaction on volunteer firefighter retention.
- *Action plan development.* Based on the survey findings, develop an action plan to address areas of concern and enhance overall volunteer satisfaction.
- *Communication of findings.* Share the survey findings with volunteer firefighters and department leadership to foster transparency and demonstrate a commitment to addressing concerns.
- *Implementation of improvements.* Actively implement the action plan and improvements based on the survey feedback to show that volunteer satisfaction is valued and taken seriously.
- *Follow-up surveys.* Conduct follow-up surveys after implementing changes to gauge the effectiveness of the improvements and whether satisfaction has increased.

- *Continuous feedback loop.* Create a continuous feedback loop by regularly seeking input from volunteer firefighters and incorporating their suggestions into ongoing improvements.

Tracking Family Satisfaction

Tracking the satisfaction of family members of volunteer firefighters is essential to understand their experiences and ensure they feel supported and valued within the fire department community.

STRATEGY STOP

Track the satisfaction of volunteers' families.
By tracking the satisfaction of family members of volunteer firefighters, you can identify areas for improvement, ensure families feel valued, and create a supportive environment that acknowledges their essential role in supporting the firefighters and the fire department community. The following strategies can help you to track the satisfaction of volunteers' families:

- *Family satisfaction survey.* Develop a family satisfaction survey that includes questions about their overall experience, communication with the fire department, available support, and any concerns they may have.
- *Anonymous and confidential.* Assure family members that their responses will remain anonymous and confidential, promoting honest feedback.
- *Survey distribution.* Distribute the family satisfaction survey through various channels, such as email, physical mail, or online platforms, to reach as many family members as possible.
- *Timing of survey.* Consider conducting the survey annually or biannually to track changes in family satisfaction over time.
- *Open-ended questions.* Alongside multiple-choice questions, include open-ended questions that allow family members to provide detailed feedback and suggestions.
- *Feedback from family events.* Gather feedback during family events or gatherings to capture real-time impressions and experiences.
- *Data analysis.* Collect (aggregate) and analyze the survey data to identify patterns and areas that may affect family satisfaction.

- *Data segmentation.* To understand any variations in satisfaction levels, segment the data by different family roles (e.g., spouses, children) and the length of time the firefighter has served.
- *Consideration of firefighter feedback.* Incorporate feedback from volunteer firefighters on how to improve family support and communication.
- *Family support meetings.* Conduct periodic family support meetings to address concerns, provide updates, and facilitate open dialogue with family members.
- *Implementation of improvements.* Based on the survey findings, develop an action plan to address areas of concern and enhance family satisfaction.
- *Communication of findings.* Share the survey results with family members and the fire department's leadership to demonstrate a commitment to improving family support.
- *Follow-up surveys.* Conduct follow-up surveys after implementing changes to gauge the effectiveness of improvements and whether family satisfaction has increased.
- *Family liaison.* Designate a family liaison within the fire department to serve as a point of contact for family members, addressing their questions and concerns.
- *Continuous feedback.* Encourage ongoing feedback from family members through regular communication channels, such as email, phone calls, or suggestion boxes.

Review Questions

1. Tracking the progress of recruitment and retention efforts is crucial for evaluating _____ and to make necessary adjustments.
2. Regularly tracking the progress of your retention plan will help you to enhance firefighter retention in your volunteer fire department how (provide at least two answers)?
3. By tracking the satisfaction of family members of volunteer firefighters, you can acknowledge their essential role in supporting the firefighters and the fire department community how (provide at least two answers)?

Discussion Questions

1. How are you currently tracking progress of your recruitment and retention plans?
2. How are you currently tracking family satisfaction?
3. What methods can you use to track progress at providing job satisfaction to your volunteers?

19

FINAL THOUGHTS

As a volunteer firefighter and as a citizen in a community protected by a volunteer fire department, I have a vested interest in the retention of volunteer firefighters. Conducting a doctoral study on strategies to retain volunteer firefighters was challenging and rewarding. The interviews that I conducted with fire service leaders were exciting and informative. The process of completing this study increased my level of knowledge and understanding of retention strategies. Since conducting this study, I have continued to research best practices from successful volunteer fire service leaders.

I encourage you to keep learning, asking questions, and moving your recruitment and retention efforts forward. Bear in mind that the responsibility for engaging and retaining volunteers in the fire service falls primarily on the shoulders of you, the individual reader. You are an active agent in the volunteer ecosystem. Become a catalyst for positive change and consider it part of your job to effect change in the volunteer fire service of the future.

Finally, I would love to hear how these strategies have made a positive impact within your community. As we are now both stakeholders in this knowledge, I also welcome learning about effective strategies you have found that were not discussed in this book. Please share your feedback and ideas with me via email to mycdevelopment@gmail.com.

ANSWERS TO CHAPTER REVIEW QUESTIONS

Chapter 1

1. What is data-driven storytelling?

   Answer: The process of transforming analytical data into visual forms that can influence the decisions of stakeholders.

2. List at least three benefits of data-driven storytelling.

   Answers: Highlight impact; show growth and need; personalize stories; visualize success; address concerns; demonstrate transparency; appeal to emotions; incorporate social media; and track volunteer progress.

3. Compelling data storytelling should include visuals that represent the _____ supporting your point.

   Answer: Evidence.

Chapter 2

1. What does history provide fire service leaders with?

   Answer: Valuable insights, knowledge, and lessons that can inform and shape the planning of the future.

2. The volunteer was started in _____, when _____ founded the Union Fire Company in Philadelphia.

   Answer: 1736; Ben Franklin.

3. The NVFC was founded in 1976 to serve as what?

   Answer: A voice for the volunteer firefighter at a national level.

Chapter 3

1. Name at least three common misconceptions about volunteer firefighters.

    Answers: Lack of training; volunteers are less skilled; volunteering is only for the young; volunteers are only needed for fires; lack of professionalism; volunteers are less equipped; volunteering is a short-term commitment; only men can be volunteers; volunteers don't respond quickly; and some people assume that volunteer firefighters might take longer to respond to emergencies because of their off-duty status.

2. What two steps are involved in identifying objectives to create an effective volunteer data–tracking plan?

    Answer: Clearly define the objectives of tracking volunteer data; determine what information you need to collect and how it will be used to support your organization's goals.

Chapter 4

1. Identify community engagement opportunities that your department can attend to _____ about what your fire service organization does and the impact that your volunteers have on your community.

    Answer: Educate the community.

2. What is a canned presentation?

    Answer: A standard presentation that an organization uses to present information systematically and consistently.

Chapter 5

1. _____ is a trend among Americans after the pandemic, in which individuals leave organizations where they don't feel valued or where they feel the organization harms their mental or physical health or interferes with social and family time.

    Answer: The Great Resignation.

2. Women volunteer in their communities _____ times more than men do.

   Answer: Three.

3. Fire service leaders should schedule regular one-on-one check-ins with volunteers to _____.

   Answers: Discuss their experiences; address any concerns; and provide opportunities for feedback.

Chapter 6

1. _____ theory states that leaders can maximize how an organization performs by meeting the needs of those with a stake in the future of the organization.

   Answer: Stakeholder.

2. For trust to occur among stakeholders, there must be _____ trust and trustworthiness.

   Answer: Leadership.

3. The five-step process for creating trust among stakeholders is called the _____.

   Answer: ELFEC or Trust creation process.

Chapter 7

1. Short sleep duration is defined as getting less than _____ hours of sleep within a typical 24-hour period.

   Answer: 7.

2. The Help Network of Northeast Ohio recommends this easy-to-remember mnemonic to remember the signs of suicide: _____.

   Answer: IS PATH WARM?

3. Generation Z are said to have a _____ attitude and are passionate about creating social change.

   Answer: Give-back.

Chapter 8

1. Leading an organization through change takes _____ and _____ from organizational leadership.

   Answer: Strategic planning; commitment.

2. What are the four reasons an organization would need to implement change?

   Answer: Structure, tasks, technology, and people changes.

3. You should have a _____ to ensure that everyone who is impacted is kept informed and allowed to provide feedback.

   Answer: Two-way communication plan.

Chapter 9

1. What are the nine critical elements that should be included in your R&R game plan?

   Answer: What, who, current resources, needed resources, potential barriers, overcoming barriers, communication, timeline, and progress check.

2. What are the three potential barriers to implementing your R&R game plan? (Hint: They all that start with the letter P.)

   Answer: People, policy, and pitfalls.

Chapter 10

1. What strategy can you use to identify and document the roles that exist within your organization—both emergency and nonemergency roles?

   Answer: Hold a brainstorming session with your members to identify and document the roles within your organization.

2. What information should job descriptions include?

    Answer: Brief overview of the role being filled, how it supports the mission of the department, list of key responsibilities, training requirements, qualifications, and commitment expectations.

3. Name at least one of the biggest mistakes that occurs during onboarding.

    Answers: Fire service leaders don't provide all the information a new member will need to be successful; or, fire service leaders hand the new member a packet of documents without explaining each document.

Chapter 11

1. What toolkit can be used to educate employers and employees on discrimination, harassment, and retaliation?

    Answer: The Fire Service Discrimination & Harassment Toolkit, from NVFC and Women in Fire.

2. How can the national Fire Service Code of Ethics be used when an internal stakeholder makes a mistake?

    Answer: If a member makes a mistake that is newsworthy, the chief can cite and provide a copy of the code to media, while refraining from commenting on a pending investigation.

3. How many search engine results do you get when you Google "fire chief arrested for theft"?

    Answer: Over seven million.

Chapter 12

1. What type of checks should be conducted of volunteers staffing camps?

    Answer: Conduct necessary background checks for all adult volunteers involved with the youth camp through the Bureau of Criminal Investigation and screen volunteers through the National Sex Offender Public Website (http://www.nsopw.gov/).

2. How should you document events your department hosts or participate in?

   Answer: Take photos and record videos for social media or the department's website.

3. What organizations should you consult for food safety guidance?

   Answer: The U.S. Department of Agriculture and your local health department.

Chapter 13

1. Good outreach plans have _____ and _____.
   Answer: Goals and objectives.

2. Objectives answer these types of questions.
   Answer: Who, what, when, why, and to what standard.

3. Effective outreach starts with _____.
   Answer: Selecting the right tool.

Chapter 14

1. How many social media users were there worldwide in 2021?
   Answer: Over 4.26 billion.

2. What is a social media content calendar?
   Answer: A collection of upcoming posts organized by date and time.

3. OpenAI's ChatGPT is a free chatbot technology that has the ability to generate what?

   Answer: Conversational, humanlike responses.

Chapter 15

1. A fire service family mentor program works with families to bring _____ to help a volunteer's social network to navigate fire service life.

   Answer: Support, guidance, and instruction.

2. Forcing engagement on family members may lead to what?

   Answers: Higher levels of dissatisfaction; less support to the volunteer.

3. Who does the fire department contact within 24 hours of a line of duty death?

   Answer: The National Fallen Firefighters Foundation via the 24-hour line-of-duty death hotline at 866-736-5868.

Chapter 16

1. Aligning nonwage benefits with the wants and needs of younger generations organization can achieve what?

   Answer: Reduce turnover among volunteers.

2. What is one limitation to creating a fun environment for volunteer retention?

   Answer: Older generations with limited financial resources obtain higher motivation from stipends than from environmental incentives.

3. Professional development opportunities do not have to be fire service focused, but can also include things that _____.

   Answer: Benefit the department.

Chapter 17

1. Recognizing and celebrating volunteer accomplishments and special occasions leads to a shared _____ among volunteers and an increase in job satisfaction.

   Answer: Sense of responsibility.

2. A LOSAP provides _____ to active volunteer members of an emergency service organization.

   Answer: Tax-deferred income benefits.

Chapter 18

1. Tracking the progress of recruitment and retention efforts is crucial for evaluating _____ and to make necessary adjustments.

   Answer: Effectiveness.

2. Regularly tracking the progress of your retention plan will help you to enhance firefighter retention in your volunteer fire department how (provide at least two answers)?

   Answers: Identify strengths and weaknesses; measure the plan's impact; and make data-driven decisions.

3. By tracking the satisfaction of family members of volunteer firefighters, you can acknowledge their essential role in supporting the firefighters and the fire department community how (provide at least two answers)?

   Answers: Identify areas for improvement; ensure volunteers feel valued; and create a supportive environment.

REFERENCES

1. Using Data for Recruitment and Retention Efforts

1. "National Fire Department Registry Quick Facts," U.S. Fire Administration, updated July 2, 2024, https://apps.usfa.fema.gov/registry/summary.
2. "Volunteer Fire Service Fact Sheet," National Volunteer Fire Council, https://www.nvfc.org/wp-content/uploads/2024/03/fire-service-fact-sheet-updated-032024.pdf.
3. "Secondary Traumatic Stress," Administration for Children & Families, accessed July 3, 2024, https://www.acf.hhs.gov/trauma-toolkit/secondary-traumatic-stress.
4. "CASA/GAL Volunteers Bring Stability while Advocating for Youth in Foster Care," *OJJDP News @ a Glance,* May/June 2022, https://ojjdp.ojp.gov/newsletter/ojjdp-news-glance-mayjune-2022/casagal-volunteers-bring-stability-while-advocating-youth-foster-care.
5. Salwa Mohamed El Habib, "Perceptions of Court Appointed Special Advocates on Volunteer Turnover" (PhD diss., Walden University, 2019), https://scholarworks.waldenu.edu/cgi/viewcontent.cgi?article=8171&context=dissertations.
6. National Fire Protection Agency. 2022. U.S. Fire Department Profile 2021. NFPA. Accessed March 19, 2025. https://www.emergent.tech/blog/firefighters-called-out#:~:text=According%20to%20research%20from%20the,a%20wide%20variety%20of%20tasks.
7. "Task Force Report & Recommendations," Ohio Task Force on Volunteer Fire Service, January 5, 2023, https://com.ohio.gov/static/documents/TaskForceReportFinal.pdf.
8. "The Fire Service in the United States of America," National Volunteer Fire Council, accessed July 3, 2024, https://www.nvfc.org/wp-content/uploads/2023/01/2023-Fire-Service-Infographic.pdf.

2. Understanding the History of the Volunteer Fire Service and Fire Department

1. Craig Collins, "The Heritage and Evolution of America's Volunteer Fire Service," in *A Proud Tradition: 275 Years of the American Volunteer Fire Service* (Tampa, FL: National Volunteer Fire Council, 2012), https://www.nvfc.org/wp-content/uploads/2015/10/Anniversary_Publication.pdf.
2. "Articles of the Union Fire Company, 7 December 1736," National Archives: Founders Online, accessed July 3, 2024, https://founders.archives.gov/documents/Franklin/01-02-02-0024.
3. Olivia Kozlevcar, "Red Cross Volunteers Who Helped America Heal after 9/11 Continue to Deliver Hope," American Red Cross, September 9, 2022, https://www.redcross.org/local/new-york/greater-new-york/about-us/news-and-events/news/volunteers-who-helped-america-heal.html.
4. U.S. Fire Administration, *2023 U.S. Fire Administrator's Summit on Fire Prevention and Control, Workgroup Report*, October 2022–August 2023, https://www.firehero.org/wp-content/uploads/2024/01/usfa-summit-workgroup-report-011224.pdf.
5. The Center for Fire, Rescue, and EMS Health Research, NDRI-USA, accessed July 3, 2024, https://www.ndri-usa.org/centers/the-center-for-fire-rescue-and-ems-health-research.
6. "About the National Firefighter Registry (NFR) for Cancer," National Institute for Occupational Safety and Health (NIOSH), accessed July 3, 2024, https://www.cdc.gov/niosh/firefighters/registry/aboutnfr.html.
7. Firefighter Life Safety Initiatives, Everyone Goes Home, accessed July 3, 2024, https://www.everyonegoeshome.com/16-initiatives/.
8. S. A. Jahnke, et al., "The Prevalence and Health Impacts of Frequent Work Discrimination and Harassment among Women Firefighters in the US Fire Service," *BioMed Research International 2019*, March 20, 2019. https://www.ncbi.nlm.nih.gov/pmc/articles/PMC6446094/.
9. Kim Quiros, "The National Volunteer Fire Council," in *A Proud Tradition*. https://www.nvfc.org/wp-content/uploads/2015/10/Anniversary_Publication.pdf.
10. "Mission Statement of the Emergency Responder Safety Institute," ResponderSafety.com, accessed July 3, 2022, https://www.respondersafety.com/about-us/.

3. The Value of the Volunteer

1. "Introduction," *Successful Strategies for Recruiting, Training, and Utilizing Volunteers: A Guide for Faith- and Community-Based Service Providers*, Substance Abuse and Mental Health Services Administration, U.S. Department of Health and Human Services, https://www.samhsa.gov/sites/default/files/volunteer_handbook.pdf.
2. "NSW Multicultural Volunteering Report," The Centre for Volunteering, https://www.volunteering.com.au/wp-content/uploads/2022/12/NSW-Multicultural-Volunteering-Report-2022.pdf

3. "Disaster Volunteer Opportunities," American Red Cross, https://www.redcross.org/volunteer/volunteer-opportunities/disaster-volunteer.html#:~:text=Volunteers%20constitute%20about%2090%20percent,them%20home%20and%20apartment%20fires.
4. "Fire Department Calls," National Fire Protection Association, updated September 2022, https://www.nfpa.org/education-and-research/research/nfpa-research/fire-statistical-reports/fire-department-calls?l=67.
5. "The Fire Service in the United States of America," National Volunteer Fire Council, accessed July 3, 2024, https://www.nvfc.org/wp-content/uploads/2023/01/2023-Fire-Service-Infographic.pdf.
6. "Value of Volunteer Time: Methodology," Independent Sector, accessed July 2, 2024, https://independentsector.org/value-volunteer-time-methodology/.
7. "Occupational Outlook Handbook," U.S. Bureau of Labor Statistics, last modified April 17, 2024, https://www.bls.gov/ooh/.
8. "Value of Volunteer Time," Independent Sector, April 23, 2024, https://independentsector.org/resource/value-of-volunteer-time/.
9. "Guidance on the Protection of Personal Identifiable Information," U.S. Department of Labor, accessed June 13, 2024, https://www.dol.gov/general/ppii.

5. Understanding the Current Problem with the U.S. Volunteer Fire Service

1. Candice McDonald, "Retention of Internal Stakeholders in the U.S. Volunteer Fire Service" (DBA diss., Walden University, 2016), https://scholarworks.waldenu.edu/dissertations/3180/.
2. "Volunteer Fire Service Fact Sheet," National Volunteer Fire Council, https://www.nvfc.org/wp-content/uploads/2024/03/fire-service-fact-sheet-updated-032024.pdf.
3. "Spotlighting the CEV Series: Formal Volunteering and Informal Helping in America," AmeriCorps, accessed July 3, 2024, https://americorps.gov/sites/default/files/document/CEV%20Volunteering%20Helping_030123_final_508.pdf.
4. "Volunteering and Civic Life in America," AmeriCorps, accessed July 3, 2024, https://americorps.gov/about/our-impact/volunteering-civic-life.
5. "Volunteering and Civic Life in America: Research Summary," AmeriCorps, accessed July 3, 2024, https://americorps.gov/sites/default/files/document/volunteering-civic-life-america-research-summary.pdf.
6. *Retention and Recruitment for the Volunteer Emergency Services: Challenges and Solutions*, U.S. Fire Administration, 2007, https://www.usfa.fema.gov/downloads/pdf/publications/fa-310.pdf.
7. *Critical Health and Safety Issues in the Volunteer Fire Service*, U.S. Fire Administration, December 2016, https://www.usfa.fema.gov/downloads/pdf/publications/critical_health_and_safety_issues.pdf.
8. "Demographics," AmeriCorps, accessed July 3, 2022, https://americorps.gov/about/our-impact/volunteering-civic-life/demographics.
9. "Volunteering in the United States, 2015," U.S. Bureau of Labor Statistics, February 25, 2016, https://www.bls.gov/news.release/volun.nr0.htm.
10. McDonald, "Retention of Internal Stakeholders in the U.S. Volunteer Fire Service."

11. Rita Fahy, Ben Evarts, and Gary P. Stein. "U.S. Fire Department Profile," National Fire Protection Association, August 31, 2022, https://www.nfpa.org/News-and-Research/Data-research-and-tools/Emergency-Responders/US-fire-department-profile.
12. Hubert Janicki, "How the Covid-19 Pandemic Prompted More People to Change Jobs," United States Census Bureau, May 13, 2024, https://www.census.gov/library/stories/2024/05/great-reshuffling.html
13. Taylor Telford, "'Quiet Quitting' Isn't Really about Quitting: Here Are the Signs," *Washington Post*, August 21, 2022, https://www.washingtonpost.com/business/2022/08/21/quiet-quitting-what-to-know/.
14. "2024 Gen Z and Millennial Survey: Living and Working with Purpose in a Transforming World," Deloitte, accessed July 3, 2024, https://www.deloitte.com/global/en/issues/work/content/genz-millennialsurvey.html.
15. Jim Harter, "Is Quiet Quitting Real?" Gallup, updated May 17, 2023, https://www.gallup.com/workplace/398306/quiet-quitting-real.aspx.

6. Focus on Your Stakeholders

1. Dorie Clark, "Transparency Is the New Leadership Imperative," *Harvard Business Review*, April 11, 2012, https://hbr.org/2012/04/transparency-is-the-new-leader.
2. G. T. Doran, "There's a S.M.A.R.T. Way to Write Management's Goals and Objectives." *Management Review (AMA FORUM)* 70, no. 11 (1981): 35–36.

7. Understanding Why Firefighters Disengage

1. "Volunteering in the United States, 2015," U.S. Bureau of Labor Statistics, February 25, 2016, https://www.bls.gov/news.release/volun.nr0.htm.
2. Candice McDonald, "Retention of Internal Stakeholders in the U.S. Volunteer Fire Service" (DBA diss., Walden University, 2016), https://scholarworks.waldenu.edu/dissertations/3180/.
3. "Code Audit," Canadian Code for Volunteer Involvement, https://audit.volunteer.ca/node/74
4. "What Are Sleep Deprivation and Deficiency?" National Heart, Lung, and Blood Institute, updated on March 24, 2022, https://www.nhlbi.nih.gov/health/sleep-deprivation.
5. "Sleep," The Nutrition Source, Harvard T. H. Chan School of Public Health, https://www.hsph.harvard.edu/nutritionsource/sleep/.
6. "Shift Work Sleep Disorder (SWSD)," Cleveland Clinic, last reviewed April 27, 2023, https://my.clevelandclinic.org/health/diseases/12146-shift-work-sleep-disorder.
7. "NIOSH Training for Nurses on Shift Work and Long Work Hours," National Institute for Occupational Safety and Health, accessed July 3, 2022, https://www.cdc.gov/niosh/work-hour-training-for-nurses/longhours/mod7/03.html.
8. G. Hofer-Tinguely, et al., "Sleep Inertia: Performance Changes after Sleep, Rest and Active Waking. *Cognitive Brain Research* 22, no. 3 (2005): 323–31.

9. Taylor Shockey, "Short Sleep Duration by Occupation Group," NIOSH Science Blog, Centers for Disease Control and Prevention, March 6, 2017, https://blogs.cdc.gov/niosh-science-blog/2017/03/06/sleep-by-occupation/.
10. René E. Cornier, "Sleep Disturbances," chap. 77 in *Clinical Methods: The History, Physical, and Laboratory Examinations, 3rd ed* (Butterworths, 1990), https://www.ncbi.nlm.nih.gov/books/NBK401/
11. McDonald, "Retention of Internal Stakeholders in the U.S. Volunteer Fire Service."
12. "Sleep," Centers for Disease Control and Prevention," accessed July 3, 2024, https://www.cdc.gov/sleep/index.html.
13. "What Are Sleep Deprivation and Deficiency?" National Heart, Lung, and Blood Institute, updated on March 24, 2022, https://www.nhlbi.nih.gov/health/sleep-deprivation
14. https://ihpm.org/wp-content/uploads/2020/05/CDC_Survey-1.pdf.
15. "Sleep Diary," National Health, Lung, and Blood Institute, January 2019, https://www.nhlbi.nih.gov/resources/sleep-diary.
16. "Sleep Brochure," National Health, Lung, and Blood Institute, December 2018, https://www.nhlbi.nih.gov/resources/sleep-brochure.
17. https://ihpm.org/wp-content/uploads/2020/05/CDC_Survey-1.pdf.
18. E. H. Jang et al., "The Development of a Sleep Intervention for Firefighters: The FIT-IN (Firefighter's Therapy for Insomnia and Nightmares) Study," *International Journal of Environmental Research and Public Health* 17, no. 23 (2020): 8738, https://www.ncbi.nlm.nih.gov/pmc/articles/PMC7727785/.
19. "Supporting Women in Fire and EMS," U.S. Fire Administration, https://www.usfa.fema.gov/a-z/supporting-women-in-fire-and-ems/.
20. R. E. Cash et al., "Comparison of Volunteer and Paid EMS Professionals in the United States," *Prehospital Emergency Care* 25, no. 2 (2021): 205–12, https://pubmed.ncbi.nlm.nih.gov/32271639/.
21. "Employment," Advocates for Trans Equality, https://transequality.org/issues/employment.
22. "How Many Adults and Youth Identify as Transgender in the United States?" The Williams Institute, University of California, Los Angeles School of Law, June 2022, https://williamsinstitute.law.ucla.edu/publications/trans-adults-united-states/.
23. L. Smart Richman L and M. R. Leary, "Reactions to Discrimination, Stigmatization, Ostracism, and Other Forms of Interpersonal Rejection: A Multimotive Model," *Psychological Review* 116, no. 2 (2009): 365–83, https://www.ncbi.nlm.nih.gov/pmc/articles/PMC2763620/.
24. McDonald, "Retention of Internal Stakeholders in the U.S. Volunteer Fire Service."
25. 2021 National Emergency Medical Services Education Standards (Washington, DC: National Highway Traffic Safety Administration, 2021), https://www.ems.gov/assets/EMS_Education-Standards_2021_FNL.pdf.
26. "Mental Illness," National Institute of Mental Health, updated March 2023, https://www.nimh.nih.gov/health/statistics/mental-illness.
27. "First Responders: Behavioral Health Concerns, Emergency Response, and Trauma," *Disaster Technical Assistance Center Supplemental Research Bulletin.* Substance Abuse and Mental Health Services Administration, May 2018, https://www.samhsa.gov/sites/default/files/dtac/supplementalresearchbulletin-firstresponders-may2018.pdf.

28. "Mental Health in the Workplace," Workplace Health Promotion, Centers for Disease Control and Prevention, accessed July 3, 2024, https://www.cdc.gov/workplacehealthpromotion/tools-resources/workplace-health/mental-health/index.html.
29. I. H. Stanley et al., "Career Prevalence and Correlates of Suicidal Thoughts and Behaviors among Firefighters," *Journal of Affective Disorders* 187 (2015): 163–71.
30. R. Z. Goetzel et al., "Mental Health in the Workplace: A Call to Action Proceedings From the Mental Health in the Workplace: Public Health Summit," *Journal of Occupational and Environmental Medicine* 60, no. 4 (2018): 322–30, https://www.ncbi.nlm.nih.gov/pmc/articles/PMC5891372/.
31. "Behavioral Health: Key Issues in Fire and Emergency Services," U.S. Fire Administration, https://www.usfa.fema.gov/about/usfa-events/2022-10-11-usfa-summit/behavioral-health/#:~:text=Issue%3A%20An%20increasing%20number%20of,emergency%20services%20to%20the%20public.
32. "Share the Load Program," National Volunteer Fire Council, accessed July 3, 2024, https://www.nvfc.org/programs/share-the-load-program.
33. "Mission Statement," Federation of Fire Chaplains, accessed July 3, 2024, https://ffc.wildapricot.org/AboutFFC/.
34. Joshua R. Rhodes et al., "Posttraumatic Growth-Oriented Peer-Based Training among U.S. Veterans: Evaluation of Post-intervention and Long-Term Follow-Up Outcomes," *Frontiers in Psychology* 14 (2024): 1322837, https://www.frontiersin.org/journals/psychology/articles/10.3389/fpsyg.2023.1322837/full.
35. "Critical Incident Stress Management," National Interagency Fire Center, accessed July 3, 2024, https://www.nifc.gov/resources/taking-care-of-our-own/about-critical-incident-stress-management.
36. "The CISM Industry International Certifications," University of Maryland, Baltimore County Emergency Health Services, accessed July 3, 2024, https://ccism-cert.org/cism/.
37. "Mental Illness," National Institute of Mental Health, https://www.nimh.nih.gov/health/statistics/mental-illness.
38. Allison Szramek, "Benefits of Therapy Dogs among College-Aged Communication Sciences and Disorders Students," Illinois State University Research and eData, November 11, 2020, https://ir.library.illinoisstate.edu/cgi/viewcontent.cgi?article=1028&context=giscsd.
39. Colleen A. Dell et al., "A Case Study of the Patient Wait Experience in an Emergency Department with Therapy Dogs." *Patient Experience Journal* 6, no. 1 (2019): 115–26, https://www.researchgate.net/publication/334072717_A_case_study_of_the_patient_wait_experience_in_an_emergency_department_with_therapy_dogs.
40. Therapy Dogs International brochure, accessed July 3, 2024, https://www.tdi-dog.org/images/GeneralBrochure.pdf.
41. "AKC Recognized Therapy Dog Organizations," American Kennel Club, accessed July 3, 2024, https://www.akc.org/sports/title-recognition-program/therapy-dog-program/therapy-dog-organizations/.
42. McDonald, "Retention of Internal Stakeholders in the U.S. Volunteer Fire Service."
43. "Comparing Characteristics and Selected Expenditures of Dual- and Single-Income Households with Children," Monthly Labor Review, U.S. Bureau of Labor Statistics,

September 2020, https://www.bls.gov/opub/mlr/2020/article/comparing-characteristics-and-selected-expenditures-of-dual-and-single-income-households-with-children.htm.
44. "Employment Characteristics of Families—2023," U.S. Bureau of Labor Statistics, April 24, 2024, https://www.bls.gov/news.release/pdf/famee.pdf.
45. J. Y. Huynh, D. Xanthopoulou, and A. H. Winefield, "Social Support Moderates the Impact of Demands on Burnout and Organizational Connectedness: A Two-Wave Study of Volunteer Firefighters. *Journal of Occupational Health Psychology 18*, no. 1 (2013), 9–15.
46. https://ems.ohio.gov/certifications/fire-service/fire-service-certificates-to-practice.
47. "National Fire Department Registry Quick Facts," U.S. Fire Administration, updated July 2, 2024, https://apps.usfa.fema.gov/registry/summary.
48. *Reminiscences of Dr. J. Fred Lembright* (1973), interview by Dr. Christopher M. King, http://local.rodmanlibrary.com/transcripts/lembright.pdf.
49. Rebecca E. Cash et al. "Comparison of Volunteer and Paid EMS Professionals in the United States." *Prehospital Emergency Care* 25, no. 2 (2021): 205-212. https://doi.org/10.1080/10903127.2020.1752867
50. T. P. Møller et al., "Why and When Citizens Call for Emergency Help: An Observational Study of 211,193 Medical Emergency Calls." *Scandinavian Journal of Trauma, Resuscitation and Emergency Medicine* 23, no. 88 (2015), https://www.ncbi.nlm.nih.gov/pmc/articles/PMC4632270/.
51. "Fire Department Calls," National Fire Protection Association, https://www.nfpa.org/education-and-research/research/nfpa-research/fire-statistical-reports/fire-department-calls.
52. Charlynn Burd, Michael Burrows, and Brian McKenzie, "Travel Time to Work in the United States: 2019," American Community Survey Reports, March 2021, https://www.census.gov/content/dam/Census/library/publications/2021/acs/acs-47.pdf.
53. T. S. Rao and V. Indla, "Work, Family or Personal Life: Why Not All Three?" *Indian Journal of Psychiatry* 52, no. 4 (2010): 295–97, https://www.ncbi.nlm.nih.gov/pmc/articles/PMC3025152/.
54. "Leading a Generationally Diverse Workforce" course syllabus, U.S. Office of Personnel Management, https://www.opm.gov/policy-data-oversight/training-and-development/reference-materials/online-courses/leading-a-generationally-diverse-workforce/LeadingAGenerationallyDiverseWorkforce.pdf.
55. Bureau of Labor Statistics, "Volunteering in the United States, 2015," https://www.bls.gov/news.release/volun.nr0.htm.
56. Kim Parker and Ruth Igielnik, "On the Cusp of Adulthood and Facing an Uncertain Future: What We Know About Gen Z So Far," Pew Research Center, May 14, 2020, https://www.pewresearch.org/social-trends/2020/05/14/on-the-cusp-of-adulthood-and-facing-an-uncertain-future-what-we-know-about-gen-z-so-far-2/.
57. Cheryl Hardy, "Gen Z: The Next Generation of Nonprofit Donors and Volunteers," Convergent Nonprofit Solutions, https://www.convergentnonprofit.com/blog/next-generation-donors/.
58. NVFC Virtual Classroom, National Volunteer Fire Council, https://virtualclassroom.nvfc.org.

59. Heather Kirkland and Wendy Walsh, "Generational Perspectives in Emergency Management—a Glimpse into Understanding," FEMA Higher Education Program, April 10, 2017, https://training.fema.gov/hiedu/emtheoryresearch.aspx.
60. Office of Personnel Management, "Leading a Generationally Diverse Workforce."
61. "How to Build a Mentoring Program: A Mentoring Program Toolkit," United States Patent and Trademark Office, accessed July 3, 2024, https://www.uspto.gov/sites/default/files/documents/Mentoring%20program%20toolkit.pdf.
62. "Dimension: Organizational Culture and Climate," Child Welfare Capacity Building Collaborative, Children's Bureau, accessed July 3, 2024, https://capacity.childwelfare.gov/states/topics/cqi/organizational-capacity-guide/organizational-culture-and-climate.
63. B. Schneider, M. G. Ehrhart, and W. H. Macey, "Organizational Climate and Culture," *Annual Review of Psychology 64* (2013): 361–88, https://pubmed.ncbi.nlm.nih.gov/22856467/.
64. Velma D. Azia, "Employee Turnover Intentions in the Retail Industry," (PhD diss., Walden University, 2016), https://scholarworks.waldenu.edu/dissertations/3180/.
65. S. A. Jahnke et al., "The Health of Women in the US Fire Service." *BMC Women's Health* 12, vol. 39 (2012).
66. Volunteer Retention Research Report, National Volunteer Fire Council, August 2020, https://www.nvfc.org/wp-content/uploads/2020/08/20Aug-NVFC-Retention-Research-Report-FINAL.pdf.

8. Becoming a Catalyst for Change in a World of Resistance

1. Kotter, John. "The 8-Step Process for Leading Change." Kotter Inc. https://www.kotterinc.com/methodology/8-steps/

10. Setting the Course for Organizational Success

1. Hakikur Rahman, "Barriers to Successful Implementation of Organizational Change in the Financial Services Sector," (PhD diss., Walden University, 2017), https://scholarworks.waldenu.edu/cgi/viewcontent.cgi?article=4283&context=dissertations.
2. Lyn Ketelsen, Karen Cook, and Bekki Kennedy. *The HCAHPS Handbook 2: Tactics to Improve Quality and the Patient Experience* (Fire Starter Publishing, 2014).

11. The Power of Organizational Brand and Reputation

1. "Position Statements," U.S. Fire Administration, last reviewed January 26, 2023, https://www.usfa.fema.gov/about/position-statements/.
2. "Fire Service Reputation Management White Paper," Cumberland Valley Volunteer Firemen's Association, accessed July 2, 2024, https://firefighterbehavior.com/wp-content/uploads/2017/09/Reputation-Management-White-Paper.pdf.

3. R. N. Lipari and S. L. Van Horn, "Trends in Substance Use Disorders among Adults Aged 18 or Older," in *The CBHSQ Report,* Substance Abuse and Mental Health Services Administration, 2013, https://www.ncbi.nlm.nih.gov/pubmed/28792721/.
4. M. G. Carey et al., "Sleep Problems, Depression, Substance Use, Social Bonding, and Quality of Life in Professional Firefighters," *Journal of Occupational and Environmental Medicine 53*, no. 8 (2011): 928–33, https://www.ncbi.nlm.nih.gov/pmc/articles/PMC3486736/.
5. C. K. Haddock et al., "Alcohol Use and Problem Drinking among Women Firefighters," *Women's Health Issues* 27, no. 6 (2017): 632–38, https://www.ncbi.nlm.nih.gov/pmc/articles/PMC5694370/.
6. "Implications of Drug Use for Employers," National Safety Council, accessed July 4, 2024, https://www.nsc.org/work-safety/safety-topics/drugs-at-work/substances.
7. Michael A. Hogg, "Self-Uncertainty and Group Identification: Consequences for Social Identity, Group Behavior, Intergroup Relations, and Society," chap. 5 in *Advances in Experimental Social Psychology 64*, ed. Bertram Gawronski, (Academic Press, 2021), 263–316, https://www.sciencedirect.com/science/article/abs/pii/S0065260121000150.
8. Clayton Neighbors, Dawn W. Foster, and Nicole Fossos, "Peer Influences on Addiction," chap. 33 in *Principles of Addiction*, ed. Peter M. Miller (Academic Press, 2013), 323-31, https://doi.org/10.1016/B978-0-12-398336-7.00033-4.
9. "Opioids at Work Employer Toolkit," National Safety Council, 2019, https://www.nsc.org/getmedia/d4b0576f-ae9f-4d84-a072-f3aa89d44e03/getting-started.pdf.aspx.
10. Kathy Gurchiek, "SHRM Blue Ribbon Commission Issues Report, Strategies to End Workplace Racism," May 24, 2021, https://www.shrm.org/ResourcesAndTools/hr-topics/behavioral-competencies/global-and-cultural-effectiveness/Pages/SHRM-Blue-Ribbon-Commission-Issues-Report-Strategies-to-End-Workplace-Racism.aspx.
11. "Promising Practices for Preventing Harassment," U.S. Equal Employment Opportunity Commission, November 21, 2017, https://www.eeoc.gov/laws/guidance/promising-practices-preventing-harassment.
12. Fire Service Discrimination & Harassment Toolkit, National Volunteer Fire Council, 2023, https://www.nvfc.org/wp-content/uploads/2023/04/Discrimination-and-Harassment-Toolkit.pdf.
13. Paul Healy and George Serafeim, "How to Scandal-Proof Your Company," *Harvard Business Review*, July–August 2019, https://hbr.org/2019/07/how-to-scandal-proof-your-company.
14. DiNapoli, Thomas P., *Red Flags for Fraud*, Office of the New York State Comptroller, 2015. https://www.osc.ny.gov/files/local-government/publications/pdf/red_flags_fraud.pdf.
15. Candice McDonald, "Seven Tips for Avoiding Theft in the Firehouse," Firefighters Association of the State of New York, https://fasny.com/magazine_articles/seven-tips-avoiding-theft-firehouse/.

12. Citizen Engagement for Recruitment

1. "Community Emergency Response Team (CERT)," Federal Emergency Management Agency, https://www.fema.gov/emergency-managers/

individuals-communities/preparedness-activities-webinars/community-emergency-response-team

14. Incorporating Social Media

1. Stacy Jo Dixon, "Number of Social Media Users Worldwide from 2017 to 2028," Statista, May 17, 2024, https://www.statista.com/statistics/278414/number-of-worldwide-social-network-users/.
2. Christina Pavlou, "How to Recruit on Facebook," Workable Technology, July 2024, https://resources.workable.com/tutorial/recruit-on-facebook.
3. Jenn Chen, "Choosing the Right Social Media Channels for Your Business," Sprout Social, September 2, 2021, https://sproutsocial.com/insights/social-media-channels/.

15. Volunteering Is a Family Affair

1. *NVFC Family Guide to Health, Wellness, and Cancer Prevention, 2nd ed* (National Volunteer Fire Council, 2024), https://www.nvfc.org/wp-content/uploads/2024/06/Family-Guide-2nd-edition-online.pdf.
2. S. L. Cowlishaw, L. Evans, and J. McLennan, "Balance between Volunteer Work and Family Roles: Testing a Theoretical Model of Work–Family Conflict in the Volunteer Emergency Services. *Australian Journal of Psychology* 62, no. 3 (2010), 169–78, https://www.tandfonline.com/doi/full/10.1080/00049530903510765.
3. Gerald R. Patterson, "The Next Generation of PMTO Models," *Behavior Therapist* 28, no. 2 (2005): 25–32, https://pmto.nl/documenten/nextgeneration.pdf.
4. Candice McDonald, "Retention of Internal Stakeholders in the U.S. Volunteer Fire Service" (DBA diss., Walden University, 2016), https://scholarworks.waldenu.edu/dissertations/3180/.
5. "A Preparedness Guide for Firefighters and Their Families," National Wildfire Coordinating Group Risk Management Committee, July 2019. https://www.fs.usda.gov/sites/default/files/2020-04/preparedness_guide_for_firefighters_and_their_families.pdf.

16. The Value of Nonwage Benefits

1. Candice McDonald, "Retention of Internal Stakeholders in the U.S. Volunteer Fire Service" (DBA diss., Walden University, 2016), https://scholarworks.waldenu.edu/dissertations/3180/.
2. McDonald, "Retention of Internal Stakeholders in the U.S. Volunteer Fire Service."
3. Melanie Hanson, "Average Cost of College & Tuition," Education Data Initiative, updated May 28, 2024, https://educationdata.org/average-cost-of-college.
4. McDonald, "Retention of Internal Stakeholders in the U.S. Volunteer Fire Service."

17. Recognizing Volunteer Success

1. Velma D. Azia, "Employee Turnover Intentions in the Retail Industry" (PhD diss., Walden University, 2016), https://scholarworks.waldenu.edu/dissertations/3180/.
2. Shannon Gary Coffey, "Strategies to Increase Employees' Morale" (DBA diss., Walden University, 2021), https://scholarworks.waldenu.edu/cgi/viewcontent.cgi?article=12445&context=dissertations.
3. Bruce Linger, "Defined Benefit LOSAP Plan or Defined Contribution LOSAP Plan—That Is the Question," https://www.losap.com/p/db-vs-dc#:~:text=is%20the%20Question-,Defined%20Benefit%20LOSAP%20Plan%20or%20Defined,Plan%20%E2%80%93%20That%20is%20the%20Question&text=A%20Length%20of%20Service%20Award,%2C%20EMS%2C%20or%20rescue%20department.

INDEX

A
accountability 74, 77, 170, 183
action plan 80, 281
active listening 93
administrative procedures in onboarding 172
advocacy
  empathy in fire service leaders and 81
  national fire service organizations and 14–20, 24
AFG (Assistance to Firefighters Grant) Program 15
Alliance of Therapy Dogs 125
American Red Cross 14, 31
AmeriCorps social media toolkit 42
anxiety 105, 118, 124, 231
apprenticeship programs 141
articles of the Union Fire Company 11–13
artificial intelligence (AI)
  in marketing 222–226
  OpenAI 222
assessing
  department PPE needs 104
  during volunteer rounding 178
  organizational sleep health 97
  volunteer engagement 91–95
Assistance to Firefighters Grant (AFG) Program 15
associations
  Cumberland Valley Volunteer Firefighters Association 21, 182
  Delaware Volunteer Firefighters Association 68
  Firefighters Association of the State of New York (FASNY) 16
  Maryland State Firefighters Association (MSFA) 264
awards, state and national 273–274
awareness opportunities 195

B
baby boomers 134
background checks 199, 208
backward goal-setting session 74–75
barriers
  for female volunteers 102–103, 105
  people 164
  pitfalls 164
  policy 164
  to retention 90–91
  to specific goals 164
  to volunteering 89–90
baseline data, collecting 87
battle buddy system 251–253
Beeson, Brandon 107
behavioral health. *See also* mental health
  local partners 114–117
  peer support programs 117–122, 232
  resources 111–112, 115–117
  retention and 230
  services for volunteer families 233
  substance use and 184–187
  workplace accommodations and 122–123
benchmark checks 165
benchmarking 279, 281
benefits. *See* nonwage benefits

307

binge drinking 186
Blanchard, Ken 161
Boulder Crest Institute for Posttraumatic Growth 119
branding 179–181
  social media and 227
Bright and Beautiful Therapy Dogs 125
building trust 73
Bunker, Diane 198
Bureau of Criminal Investigation 199
burnout 1, 81, 118, 231

C

camaraderie
  activities 259, 265–267
  fostering 62, 118, 137, 195, 265
  generational value of 138
Canva 39, 41, 175
career
  development 260–261
  plans 175–176
Carter, Shelly 199
CASA (court-appointed special advocate) 2
CBT (cognitive behavioral therapy) 114
CDC (Centers for Disease Control) 96
CDP (Criteria Development Panel) 15
Center for Fire, Rescue, and EMS Health Research (CFREHR) 18
Centers for Disease Control and Prevention (CDC) 96
CERT (Community Emergency Response Team) 206–207
CFREHR (Center for Fire, Rescue, and EMS Health Research) 18
CFSI (Congressional Fire Service Institute) 15, 22
Challenges of the Firefighter Marriage 229
change
  champions of 156–158
  communication during 155–156
  leading an organization through 150
  navigating 149–150
  people 153
  process 155
  reasons for 152–153
  stakeholder input 78
  structural 152
  task 152
  technology 153
  transparency in 154
chaplains, fire service 112, 114
charting, electronic 153
ChatGPT 222
checks and balances
  financial records and inventory lists 190
  social media accounts 228
Chief, Jersey City (NJ) Fire Department x
child-care assistance 237
circadian clock 95
circadian rhythms 100–101
CISM (critical incident stress management) 122
citizen engagement programs
  "Be a Firefighter for the Day" events 197
  citizen fire academy 195–196
  Feel the Heat 197–198
  fire department open house 201–203
  listening tours 47–48
  recruitment tables 203–205
  supporting local sporting events 205
  youth fire camp 199–201
City of Hartford, Connecticut, Fire Department 199
civic engagement 219
civic service groups 48, 50
Clark, Dorie 71
code of ethics, fire service 183–184
cognitive behavioral therapy (CBT) 114
collective voice 14–15
communication. *See also* open communication; two-way communication
  customizing 207
  of expectations 62
  fact sheets 242–243
  multichannel 240
  open 81, 232, 237, 254
  planning for crisis 240
  transparent 136
  trees 253
community
  engagement 38, 47, 197, 201, 233
  events 47–48, 266

member success stories 43–44
resource library 117
Community Emergency Response Team (CERT) 206–207
compassion fatigue 1–2, 231
Comstock, Chip 17
conflict resolution 138
Congressional Fire Services Institute (CFSI) 15, 22
Congressional Fire Services Institute Excellence in Fire Service-Based EMS Award 274
consent, obtaining 228
content calendar for social media 221
continuous improvement 36, 63, 280
cookouts, potlucks, and community meals 146–147
copyright laws 228
corporate philanthropy 34
costs
  firefighter training 54
  value of volunteer time 32
court-appointed special advocate (CASA) 2
credit checks on financial managers 190
credit hour system 99–100
crisis communication plan 240
Criteria Development Panel (CDP) 15
critical incident stress management (CISM) 122
critical thinking 10–11
criticism, online 226–227
cultural sensitivity training 233–234
cultures of
  learning 57
  ownership 77
  trust 81
Cumberland Valley Volunteer Firefighters Association (CVVFA) 21–22, 182

D

data
  analysis and reporting 80
  baseline 87
  categories of 35
  demographic 88–89
  engagement 88–89
  entry procedures 35
  historical 87
  identifying objectives 34
  management systems 34–35
  qualitative 36, 87
  quantitative 36, 87
  recruiting and 3, 89–90
  secure storage 35
  segmentation 281, 282
  storytelling and 4–6
  supporting department efforts with 37–39
  tracking of volunteers 6, 34–36, 87–89, 90
data-driven approaches 3
data-driven recruiting 89
deadlines for goal completion 165
death in line-of-duty (LODD) 255–256
Delaware Volunteer Firefighters Association 68
demographics 279
departments. *See also* Fire Department of New York (FDNY); Winona (OH) Fire Department
  City of Hartford, Connecticut Fire Department 199
  East Franklin (NJ) Fire Department 99, 269
  Jersey City (NJ) Fire Department x, 259
  Union Fire Company 11
depression
  family stressors 231
  mental health and 108, 114, 118–119
  sleep deprivation and 97
DFWP (Drug-Free Workplace Policy) 186
Dill, Jeff 114
discipline policy 200
discrimination and harassment 102, 105, 182, 233
  addressing 187–188
disengagement 61–63
  researching 91
  symptoms of 61
  training demands 127
  understanding 86, 105, 125–132
  work commute requirements and 131
Doyle, Daryl 60
Drug-Free Workplace Policy (DFWP) 186

dual-income families and
    volunteering 126–127
duty crew requirement for emergency
    response 99–100

E

EAP (Employee Assistance Program) 123
East Franklin (NJ) Fire Department 99,
    269
education opportunities 195
EFNEP (Extended Food and Nutrition
    Education Program) 235–236
electronic charting 153
elevator pitch for recruitment events 204
ELFEC (Engage, Listen, Frame, Envision,
    Commit) trust creation process 71
email and phone scripts 49
embezzlement scandals in the fire
    service 189
EMDR (eye movement desensitization
    and reprocessing) 114
emergency medical service (EMS) 1
emergency medical technician
    (EMT) 130
Emergency Responder Safety Institute
    (ERSI) 21
emotional appeal 5
emotional impact and volunteer
    disengagement 86
empathy 80–82
Employee Assistance Program
    (EAP) 123
EMS (emergency medical service) 1
EMT (emergency medical
    technician) 130
Engage, Listen, Frame, Envision,
    Commit (ELFEC) trust creation
    process 71
engagement
    data 88–89
    metrics 36
    social media and 219–227
    understanding 85
ERSI (Emergency Responder Safety
    Institute) 21
establishing relationships with
    stakeholders 72
ethics, code of 183–184

evidence-based tactics 65
Excellence in Fire Service-Based EMS
    Award 274
exiting
    interviews 61, 92–94
    surveys 93
Expanded Food and Nutrition Education
    Program (EFNEP) 235–236
expectations, communication of 62
external stakeholders 27, 58
eye movement desensitization and
    reprocessing (EMDR) 114

F

fact sheets
    communication and 242–243
    fire department 41
    for open house events 202
Fair Labor Standards Act 127
family 229
    battle buddy system 251–253
    communication paths 239–240
    discussing line-of-duty deaths and
        injuries 256
    engagement 229–230, 232–234
    environments in the firehouse 236–237,
        239
    and firefighter guilt 131–132
    liaisons 283
    mentor programs 249–250
    onboarding 246–247
    recruitment 254–255
    satisfaction 282–283
    stressors 230–232
    support programs 232
Family Advisory Councils 233, 237
FASNY. *See* Firefighters Association of
    the State of New York (FASNY)
FDIC (Fire Department Instructors
    Conference) 108
FDNY. *See* Fire Department of New York
    (FDNY)
Federal Emergency Management Agency
    (FEMA) 15, 140, 206
Federation of Fire Chaplains (FFC) 114
feedback
    department leadership 280, 281
    providing 137, 138, 270

stakeholder 36, 48, 57, 61, 76–77, 79, 93, 140, 156, 177, 244, 282
Feel the Heat program 197–198
FEMA. *See* Federal Emergency Management Agency (FEMA)
FFC (Federation of Fire Chaplains) 114
financial
    line-of-duty death and injury planning 256
    record keeping 190–193
    resources 68
    stakeholders 31
    theft risk mitigation 190–193
financial managers and credit checks 190
Fire Department Instructors Conference (FDIC) 108
Fire Department of New York (FDNY) 27, 53, 211, 219
fire department open house 201–202
firefighter and family guilt 131–132
Firefighter Behavioral Health Alliance 114
Firefighters Association of the State of New York (FASNY) 16
firefighting and reproductive risk 108
fire service awards, local, state, and national 273–274
fire service chaplain 112, 114
Fire Service Code of Ethics 182–184
fire service interest groups. *See* interest groups
fire service organizations, state and national 14–25
Fire Service Reputation Management White Paper 182–183
flexibility
    leadership approaches and 138
    long-term leaves of absence and 59–60, 123
    training and meeting requirements and 57, 62–63, 136–137, 176, 233, 237, 255
    volunteer duties and 137
focused listening 73
focus groups 78
followers, social media 219
follow-up, postinteraction 205, 207–208, 216
food safety 199

forced-best-friend system 251
forecasting 87
    markers 87
Franklin, Ben 11
Freeman, Robert Edward 65–66
fringe-benefit rate 32–33
funding and grants 37–38
fundraising 47

G

Gagliano, Anne 229
Gagliano, Mike 229
game plan for recruitment and retention 161
gear fitting and volunteer success 102–104
gear-sharing programs 103–105
gender-responsive sensitivity 234
gender-specific issues 90, 101–108
generational
    awareness 132
    conflict 132–134
    definitions 134–136
    impacts 56–57
    management 136–138
    stereotypes and misconceptions 138–140
Generation X 134–135
Generation Z 135–136
Girls Future Firefighter Camp (GFFC) 199
goals 74, 211
    backward, goal-setting session 74–75
    deadline completion and 165
    setting 74–76
Graham, Yvette 235
grants and funding 37–38
Great Resignation 58–60
Green, Charles 70–71
group chat features 145, 157

H

harassment and discrimination. *See* discrimination and harassment
Harvard Business Review 71
health and wellness programs 187
health, physical 96–97, 267
Help Network of Northeast Ohio 110

Higher Education Learning Plan
    (HELP) 16
historical context
  in fire department education 23
  generational impacts and 56
historical data 87
historical marker 87
history
  of the Cumberland Valley Volunteer
      Firefighters Association 21–22
  of national fire service
      organizations 14–18, 22
  of the National Volunteer Fire
      Council 20–21
  recruitment and retention strategies
      and 9, 22
  of U.S. volunteer fire service 11–14
HootSuite 224
Howe, Andrea 70–71
human error factors 100
hybrid training models 128–129

I
IAFF. *See* International Association of
    Fire Fighters (IAFF)
identifying challenges 77
identifying success in recruitment and
    retention 10
impact stories 214
inclusivity
  language and 56
  recruitment and 140
influence, building 72–74
influential leaders 72–74
injuries in line-of-duty 255–256
injury rates 106–107
integrity 73
interest groups
  addressing systemic issues 19
  building movements 19
  collaboration and 24
  collective voice and 14–15
  empowering marginalized voices 19–20
  ensuring consistency and
      coordination 16–17
  grant and funding opportunities 25
  information exchange 24
  joining 24

  legislative impact of 18–19
  networking and collaboration with 18, 24
  public awareness and education 18
  reach and influence of 15–16
  resource pooling 16, 24
  standardization and accreditation 24
internal stakeholders 65
International Association of Arson
    Investigators 15
International Association of Fire
    Chiefs 15, 18, 22
International Association of Fire Fighters
    (IAFF) 15, 40
  IAFF Peer Support Training
      program 121
International Society of Fire Service
    Instructors 15
interviews
  exit 61, 92–94
  retention 280

J
Jahnke, Sara 20, 105, 108–109
Jersey City (NJ) Fire Department 259
job descriptions 170–171
job satisfaction 145–146
joint training opportunities 235

K
kitchen table talk 146
knowledge loss 58
Kotter, John 155

L
leaders
  attitude of 154–155
  influential 72–74
  stakeholder-focused 66
  successful xii
  trustworthy 69–70
lead tree for outreach 49–50
learning
  culture of 57
  from the past 9–10
Lee, Sarah 40
length-of-service award programs
    (LOSAPs) 272
Lewis, Dave 108

limitations of stakeholder theory 68
line-of-duty death (LODD) 255–256
line-of-duty injuries and deaths, financial planning 256
listening
 active 93
 focused 73
local behavioral health partnerships 114–117
LODD (line-of-duty death) 255–256
long-term perspective 11
LOSAP (length-of-service award program) 272
Love on a Leash 125

M

Make Me A Firefighter volunteer recruitment campaign 90
markers
 forecast 87
 historical 87
marketing
 with artificial intelligence (AI) 222–226
 public relations and 38
Marsar, Stephen 27, 53, 195, 211, 219
Maryland State Firefighters Association (MSFA) 264
mature generation 134
McCoy, Tom 103
memorandums of understanding 104
mental health. *See also* behavioral health impacts in fire service 90, 108–109
 NVFC First Responder Helpline 110
 pastoral intervention 112
 Share the Load support program 111–112
 suicide awareness and prevention 109–110
 therapy dogs 124–125
 wellness programs 267
mentoring programs 137, 141–143
messaging, positive and negative 213
metrics, social media 221
Microsoft Excel 37
Microsoft Word 39, 175
millennials 135
mindset, positive 30

misconceptions
 generational 138–140
 organizational branding 180–181
 of volunteer firefighters 27–30, 40
MSFA (Maryland State Firefighters Association) 264
mutual aid 25, 104

N

narrative 4
National Aeronautics and Space Administration (NASA) 100–101
National Association of State Fire Marshals 15
National Development and Research Institutes (NDRI-USA) 17–18
National Emergency Medical Services Education Standards 107
National Fallen Firefighters Foundation 256
National Firefighter Registry for Cancer 19
National Fire Protection Association (NFPA) 15, 55
National Heart, Lung, and Blood Institute 97
National Institute of Food and Agriculture 235–236
National Institute of Mental Health 122
National Institutes of Health 44
National Interagency Fire Center 122
National Preparedness Goal 206
National Safety Council 185
National Sex Offender Public Website 199
National Volunteer Fire Council (NVFC)
 on family and volunteering 229
 on firefighter training cost 54
 founding and mission 20–21
 NVFC Family Guide 247
 NVFC First Responder Helpline 110, 115, 248
 NVFC Psychologically Healthy Fire Department initiative 111, 121–122
 programs 6, 18, 90, 106, 188, 233
 promoting legislation 15–16, 40
 on volunteer demographics 53, 55
NDRI-USA (National Development and Research Institutes) 17–18

networking 260
newsletters 243–245
NFPA (National Fire Protection Association) 15, 55
nonemergency volunteers 28, 167
nonwage benefits
  attractive work environment 262–263
  camaraderie activities 265–267
  career advancement 260–261
  fringe benefits 54, 149
  professional development and training 263–264
  stakeholder theory and 68, 259
  tuition reimbursement 264–265
  uniforms and department-branded items 261–262
North American Fire Training Directors 15
NVFC. *See* National Volunteer Fire Council (NVFC)

O
objectives 211
obtaining consent 228
Office of Personnel Management 140
Ohio State Fire Academy 198
Ohio State University (OSU) 235
onboarding 171–175, 208, 246
Onieal, Denis x, 67
online criticism, navigating 226–227
OpenAI 222
open communication
  empathy and 81
  engaging family members through 232, 237, 254
  generational management and 136
open house events 201–203
organizations
  branding 179–181
  charts 172, 175, 199
  climate 90, 144
  commitment 171–172
  culture 90, 119, 144, 270
  expertise 58
  mood 154
  socialization 145
orientation, onboarding 145
OSU (Ohio State University) 235

outreach
  lead trees 49–50
  planning 48, 211–212
  tools 212–213
ownership, culture of 77

P
Pac-Man 150–151
pastoral intervention for mental health 112
peer
  influence and substance use 185
  recognition 270
  support programs 117–122, 232
people barriers 164
people change 153
performance evaluations 3, 279
personal identifiable information (PII) 35
personal protective equipment (PPE) 5, 15, 103
Peters, Bill 259
Pet Partners 125
Pew Research Center 135
phone and email scripts 49
phone tree 240–241, 253
photo release forms 44
physical health 96–97
PII (personal identifiable information) 35
pitfall barriers 164
PivotTable functionality 37
planning
  for social media 221
  strategic 150
playbooks 161
play-pay certificates 39
Point mobile application 36–37
policy barriers 164
poor member behavior 181–182
positive mindset 30
postinteraction follow-up 205, 207–208, 216
postresponse engagement 216–217
posttraumatic stress disorder (PTSD)
  family stressors and 230–231
  impact 108–109
  peer support and 119
  resources 112

Powell, Wayne 9
PPE. *See* personal protective equipment (PPE)
preparation for exit interviews 93
presentations 50–52
professional development
  code of ethics 183
  interest groups and 24
  volunteer retention and 62, 138, 259
Pro Football Hall of Fame 6
Psychologically Healthy Fire Departments Initiative (NVFC) 111, 121–122
PTSD. *See* posttraumatic stress disorder (PTSD)
public relations 38

Q

qualitative data 36, 87
quantitative data 36, 87
quiet quitting 61–63. *See also* exiting; volunteer turnover

R

recognition
  activities 271–272
  leadership and 81
  of patterns and trends 10
  peer 270
  professional 25
  programs 269–271
  public-facing mechanisms 273
  retention and 62
  of volunteer families 233, 237, 240, 243, 255, 272
  of volunteers 38, 57, 136, 208, 215, 267, 269, 280
recognizing patterns in recruitment and retention 10
record keeping, financial 190–193
recruitment
  citizen engagement programs 195
  data and 3, 205
  elevator pitch 204
  game plan 161–165
  goals 74
  recognizing patterns and trends 10
  tracking 277–278

using history in 9
religious sensitivity 234
reproductive risks of firefighting 108
reputation management 58, 193
researching disengagement 91
resiliency 74
resources
  allocating 3, 38
  pooling 16
respecting time of volunteers 72, 143–144
retention
  barriers to 90–91, 144
  empathy and 81
  goals 74
  increasing 118, 229, 232, 261
  professional development and 62, 138, 259
  rates 3, 279
  recognizing patterns and trends 10
  tracking 278–280

S

16 Firefighter Life Safety Initiative 19
SAE (Society of American Engineers) 21
SAFER (Staffing for Adequate Fire and Emergency Response) Grant Program 15–16
satisfaction survey 280
Schmittendorf, Tiger 132, 213
The Science Alliance 22
SDOH (social determinants of health) 185
secondary traumatic stress disorder 2, 231. *See also* compassion fatigue
substance use and 185
self-awareness 66
self-inventory 66
self-performance evaluations 178
setting goals 74–76
Share the Load support program 111–112
shift work sleep disorder (SWSD) 96
skills matching 62
sleep
  deficiency 95
  deprivation 90, 95–97, 230
  disorder 100
  disturbances 231
  health education 97

sleep (*continued*)
  inertia 96
  initiative programs 97–100
  restriction 96
S.M.A.R.T. goals. *See* Specific, Measurable, Attainable, Relevant, Time-bound (SMART) goals
social determinants of health (SDOH) 185
social identity theory 185–186
social media
  content calendar 221
  engaging 202–204, 255
  marketing 5, 44, 219–220, 244, 271
  metrics 221
  planning 221
  policies 227–228
Society of American Engineers (SAE) 21
socioeconomic sensitivity 234
Specific, Measurable, Attainable, Relevant, Time-bound (SMART) goals 75–76, 161, 162
Staffing for Adequate Fire and Emergency Response (SAFER) Grant Program 15–16
stakeholders xii, 65. *See also* feedback, stakeholders
  establishing relationships with 72
  external 27, 58
  financial 31
  internal 65
  theory 66, 68, 180, 259, 269
stereotypes
  generational and challenging 138–140
  negative 27
storytelling 4–5, 212
strategic planning 150
strategy development from volunteer turnover 86
structural change 152
Struggle Well 119
Substance Abuse and Mental Health Services Administration 115
substance use 96, 108, 112, 182, 184–187
  peer influence and 185
success stories 243, 255
suicide awareness and prevention 109–110, 114

support systems 3, 232, 252
surveying
  families of volunteers 282–283
  objectives 79
  satisfaction 280
  stakeholders 76–80, 79–80
  volunteers 62, 94–95, 178, 215, 280–281
SWSD (shift work sleep disorder) 96
symbiotic relationships 66

T

tactics, evidence-based 65
target audience 79, 212
targeting 214
target segmentation 212
task change 152
team-building activities 266
team cohesion 82
technology change 153
Tedeschi, Richard 119
theft in the firehouse, preventing 189–193
therapy dogs 124–125
Therapy Dogs International 124–125
time of volunteers, respecting 72, 143–144
time suck 143
time value calculation for volunteer firefighters 33
toolkits
  discrimination and harassment 106
  Fire Service Discrimination & Harassment Toolkit 188
  National Safety Council Opioids at Work Employer Toolkit 186
  Psychologically Healthy Fire Departments: Implementation Toolkit 112, 121–122
  social media, AmeriCorps 42
tracking
  gear usage 104
  recruitment data 205, 277–278
  retention efforts 278–280
  social media metrics 221
  volunteer data 34–36
traditionalists 134
traditions
  in fire departments 23
  preserving and updating 10

training demands and disengagement 127–128
training models, hybrid 128–129
transgender volunteers 101–102
transparency 5, 71–73, 77
transparent communication 136
trust
　building 73
　creation process 71
　culture of 81
　reciprocity 69
trustworthiness 69
Twain, Mark 149
two-way communication 57, 155–156, 240
two-way mentoring programs 141–143

U

understanding
　cultural and social context 10
　firefighter disengagement 86
　volunteer value 31–33
Union Fire Company 11–13
University of Maryland, Baltimore County Department of Emergency Services 122
U.S. Bureau of Labor Statistics 32–33
Occupational Outlook Handbook 33
U.S. Census Bureau 59, 131
U.S. Department of Agriculture 199, 235
U.S. Department of Labor 35
U.S. Equal Opportunity Commission 188
U.S. Fire Administration (USFA) 15, 106, 182
　U.S. Fire Administrator Recruitment and Retention Work Group 16

V

validation 80–82
volunteer engagement
　assessing 91–95
　control over job duties 145
　defining 85
　flexibility in training and meeting requirements 57, 62, 176
　gender-specific issues 101–107
　metrics 36
　recognition programs 270

volunteers. *See also* recognition, of volunteers; work-life-volunteer balance
　data tracking 6, 34–36, 87–89, 90
　decline of 53–54
　demographics 3, 55, 88–89, 279
　dual-income families and 126–127
　and ill-fitting personal protective equipment 102
　management 63, 136–138
　managers 34
　misconceptions 27–30
　nonemergency 28, 167
　outreach 211
　professionalism of 28–29
　respecting time 72, 143–144
　rounding 177–178
　safety 103
　satisfaction 118, 119, 144, 280–281
　time value calculation 33
　transgender 101
　value of 30–42
volunteer turnover
　conflict resolution and 63
　COVID-19 and 58–59
　emotional impact 86
　financial impact of 54–55
　generational impacts on 55–57
　harassment, discrimination, or retaliation 187
　indirect cost of 55
　indirect impact of 58
　organizational climate and culture and 144
　organizational impact of 58
　quiet quitting 61–63
　strategy development from 86
　using data to combat 87–90

W

Warrior PATHH (Progressive and Alternative Training for Helping Heroes) 119
Webb, Bill 144
wellness programs 267
Williams, Bruce 47
Winona (OH) Fire Department 60, 235, 274
Women in Fire 20, 22, 106, 188

work commute requirements and disengagement 131
work group, recruitment and retention 16
work-life balance 61, 121, 126, 149
work-life-volunteer balance 126–132
  promoting 137, 187, 233, 237
  sleep deprivation and 96
  understanding 99, 125–132

workplace accommodations 122
work-rest cycles 100–101

Y
youth fire camps 199–201

Z
Zook, Caroline Kelso 167